Evolution and Classification

Other books by the same author

The Explanation of Organic Diversity
(Clarendon Press, Oxford, 1983)

The Problems of Evolution
(Oxford University Press, 1985)

MARK RIDLEY

Evolution and Classification

The reformation of cladism

Longman
London and New York

Longman Group Limited
Longman House, Burnt Mill, Harlow
Essex CM20 2JE, England
Associated companies throughout the world

*Published in the United States of America
by Longman Inc., New York*

© Longman Group Limited 1986

All rights reserved; no part of this publication may be reproduced, stored in a retrieval system, or transmitted in any form or by any means, electronic, mechanical, photocopying, recording, or otherwise, without the prior written permission of the Publishers.

First published 1986

British Library Cataloguing in Publication Data
Ridley, Mark
  Evolution and classification: the reformation of cladism.
  I. Biology-Classification
  I. Title
  574'.012   QH83

ISBN 0-582-44497-7

Library of Congress Cataloging in Publication Data
Ridley, Mark.
  Evolution and classification

  Bibliography: p.
  Includes index.
  1. Cladistic analysis. 2. Evolution.
I. Title.
QH83. R49   1986   575   85-5249
ISBN 0-582-44497-7 (pbk.)

Set in 10/11pt Baskerville
Produced by Longman Group (FE) Limited
Printed in Hong Kong

Contents

Preface	
1 The argument	1
2 The techniques and justification of evolutionary taxonomy	19
3 The techniques and justification of phenetic taxonomy	35
4 The techniques and justification of cladism	46
5 The reformation of cladism	86
6 Classification before Darwin	98
7 Classification and the tests of evolution	115
8 The functional criterion	125
9 Ancestral groups and sister groups	138
10 Cladograms and speciation	150
11 Evolution and classification	158
12 Conclusions	163
References	167
Glossary	183
Index	193

Preface

Perhaps I can introduce the book best by telling its history. During 1981 and 1982, when I was completing an earlier study of evolution and classification, I became increasingly aware of a new school of thought, then called 'transformed cladism' (or, as John Maynard Smith suggested, 'lapsed cladism'), which appeared to contradict nearly everything I then believed about the relation of evolution and classification. I had thought – like many others – that the theory of evolution was essential to Hennig's cladistic system, in order to justify the use of his particular set of classificatory methods. And yet here was a group of taxonomists apparently arguing that Hennig's system could do without his philosophy – without the theory of evolution. I had time to take a closer look at transformed cladism early in 1983. I was soon moved to write a polemic about the new school; but then, almost immediately, I realized that this is not the kind of subject in which a short critique can hope to have any influence. Brief arguments, be they ever so destructive, seem usually to be ignored by their intended victims, or misunderstood and then replied to with irrelevancies. It is the kind of subject that has to be dealt with at length or not at all. My original polemic now contributes to Chapters 5, 8, 9, and 10 of this book.

When I was myself first trying to find out about taxonomy, and later when I have tried to teach it, I felt the need for an introductory work, of the kind that is intelligible to the beginner, and is modern and comprehensive enough to cover all the main schools (including the several modern versions of cladism). I wrote the first four chapters (especially Chapters 2 to 4) with that need in mind, as well as because the main argument of the book – that the best known system of classification requires the theory of evolution – does not make sense outside its setting in the controversy among the three main classificatory schools.

For this final version, therefore, I have three main aspirations. To the student, or any other reader who comes to the subject with little prior knowledge (or perhaps only with rusty memories of some uninspiring, frail, and scarcely audible old expert on the genitalia of flies), I hope to introduce the main schools of classificatory thought and the modern controversy; uncommitted biologists I hope to convince of the superiority of Hennig's phylogenetic system of classification to all its competitors; and, to any biologists who may be inclined to flirt with the doctrines of transformed cladism, I hope to show that cladism really cannot be separated from the theory of evolution.

I shall not give any more details about the book's contents here, because the second half of Chapter 1 summarizes the entire argument. But I do have two things to say to the experts. One concerns that summary in Chapter 1. It is only a bare summary of what is to be argued later; it is not supposed to be convincing by itself. I have included it because the argument of the book is quite long, and readers may later on be glad to know where they are within the whole. So please do not let the brief assertions of Chapter 1 frustrate you: wait until you have seen the extended argument, which alone is meant to be persuasive. My second point concerns a subject I have not discussed in the text: the proper use of technical terms. Philosophers of classification, I know, are peculiarly given to squabbling about how terms such as evolution, phylogeny, monophyly, or cladism, should be used. But I have been writing more with an audience of general biologists than of professional taxonomists in mind. I do not think they should be much interested in verbal matters. And, in any case, they are probably unfamiliar with verbal skirmishes, customarily confined as they are to bad-tempered, often condescending digressions, and abrupt footnotes. I have, as I say, ignored all the verbal controversy, and have aimed instead only to use terms clearly and consistently. As each term comes up, I have merely said how I shall use it, although I have usually spelled out its other meanings as well (and I have added a glossary which should clear up any real ambiguities). My policy follows from the nature of the book, together with a general impatience I have with arguments about definition, and not because I am naive about the methods of taxonomic controversy. I know how each allegiance betrays itself in a name. My own choices could not help preferring one party or another and – I suppose because I have tried to pick the most common (but not universal!) usages – I have tended to favour the 'evolutionary' school. Thus, I have called Hennig's system 'cladism', and not 'phylogenetic systematics' – the term that so vexed Mayr even though others (including me) find it unusually apt. But the evolutionary taxonomist may find plenty to complain about too, such as the way I have used the term 'monophyly'. It is impossible to satisfy everyone, and I have not tried to. I have only tried to be consistent and clear and, then, as orthodox as possible. I should be grateful if my critics remember what my aim has been.

I have discussed the arguments in this book with many people, to all of whom I am grateful; and I am especially grateful to Richard Dawkins, Joseph Felsenstein, Alan Grafen, Paul Harvey, and David Hull, who have kindly read and commented on my manuscripts. The book was started while I was the Haywood Junior Research Fellow of Oriel College, Oxford; and it was completed after I had moved to become the Astor Junior Research Fellow of New College, Oxford. To these two colleges and their benefactors I am most grateful.

Mark Ridley

Oxford
February 1985

1
The argument

As the theory of evolution, in the last two decades, has enjoyed a great infusion of fundamental and far-reaching ideas, the philosophy of classification has not escaped re-examination. It is now an opportune moment to look again at the proper relation of classification and evolution, not least because (if I am correct) the most influential new idea in this area is utterly mistaken. Because the various schools of classification stand in different relations to the theory of evolution, we can consider the question by a discussion of the schools and their rival merits. In order to discuss the schools, we must first be introduced to them; and if the discussion is to be of any interest it must come to a positive conclusion, for if it did not, we should merely have replaced the problem of ignorance by one of confusion. In this work, therefore, I have three main aims, which I shall attempt roughly in order: to introduce the various schools of classification; to examine their rival merits; and (in judging them) to discover the proper relation of evolution and classification. It is an introduction to, and a work of, taxonomic controversy.

That there should be any controversy at all might, to begin with, seem strange. The principles of classification, considered superficially, might seem too straightforward to be controversial: you simply have to define groups by taxonomic characters (where 'we understand by taxonomic character any attribute of an organism (or better, of any group of organisms) by which it may differ from other organisms', Mayr 1942, p. 20). And if taxonomists are observed superficially, that is what they appear to do. They order living forms by arranging them in groups according to shared characters. More and more inclusive groups are defined by more and more generally shared characters, to result in the customary hierarchy of species contained in genera, families, orders, classes, phyla. Any form of life is classified by its characters: horses, for instance, are chordates because at some stage in their lives they possess a hollow dorsal nerve chord, segmented muscles, and a notochord; mammals because they are homeothermic

Evolution and Classification

and lactate; ungulates because (if we confine ourselves to modern forms) they are hoofed; and perissodactyls because they support their weight through the centre digit of their feet.

If that was all there were to it, classification would not be controversial. Nor, for that matter, would it be of much interest to the evolutionary biologist; it would be an intellectually humble, practical business. But classification is controversial. It is controversial because different characters define different groups, which means that taxonomists cannot both naively define groups and agree with each other. The disagreement of characters is both the fundamental source of all taxonomic controversy and the reason why there is more to classification than simple definition of groups. Any character will define a group; but different characters do not all define the same ones. For instance, the possession of a hollow dorsal nerve chord defines the chordates; but some other character, such as the possession of eyes, defines a quite different and conflicting (and unnamed) group, made up of a miscellany of invertebrates and nearly all vertebrates. So taxonomists are faced with a choice. They have to choose some characters rather than others. They might, of course, pick characters arbitrarily and define arbitrary groups, but in practice they do not. They generally try to choose particular kinds of characters; and their choice is dictated by some taxonomic principle. In this work, I shall call a principle of character choice a 'justification' (of why that character was chosen) or a 'philosophy'. Taxonomists have been forced to develop a philosophy, in addition to the practical task of defining groups, by the fact that different characters disagree. The need, however, has been supplied by more than one principle; and it is from this plurality that taxonomic controversy stems. The difference between the main taxonomic schools results from their different principles of character choice: different principles select different characters: different characters define different classifications.

The problem is magnified by the enormous quantity of characters. Every specimen possesses an infinity of characters. It is not just that there are legs, arms, eyes, and noses, together with the array of biochemical, chromosomal, and behavioural characters used by modern taxonomists (Goto 1982). These characters in turn can be described in many ways, by length, colour, or hardness; divided in many ways, into top half of leg or top quarter of leg; they can be combined in new characters, such as arm–leg or finger–rib; and these combinations can be made by any kind of weighting, as arm–leg = 1/2 (length of arm) + 1/4 (length of leg), or any other combination, such as 7/8 (length of leg) + 3/5 (length of arm). These characters are merely crude descriptions. Characters can also be defined theoretically. They can be divided into evolutionarily stable characters, which (we infer) do not often change during evolution, and evolutionarily

The argument

labile ones, which change more often. Or they can be divided into evolutionarily 'ancestral' characters and evolutionarily derived ones, which are the earlier and later evolutionary states of the same characters; fins and limbs are an example. Kinds of characters can be multiplied indefinitely. Taxonomists, however, cannot describe all of the infinite number of characters of an organism. They can only in practice work on a limited number; and even if they could study all characters they would still have to choose which ones to define groups with, because different characters define different groups. A choice has to be made. Classification is impossible without it.

How, then, are the characters to be chosen? I shall call the two main possibilities 'objective' and 'subjective' character choice. By subjective character choice, I mean that the characters used to define groups are picked arbitrarily (or subjectively), unguided by any principle. Subjective classification is undoubtedly feasible: groups can be defined by any character, and can therefore be defined by subjectively chosen characters as well as by any others. I shall call the classifications produced by subjective character choice subjective classifications and the systems that recommend such classifications subjective systems or schools.

The obvious danger is instability. If characters are chosen subjectively, different taxonomists classifying the same specimens may choose different characters, and divide them into different groups. Instability, however, is not unavoidable. Consistency can be imposed by agreement. If all taxonomists agree to choose the same kind of characters then their classifications will be stable even if the choice itself was arbitrary. For stability, only agreement among taxonomists is needed, not necessarily reasoned agreement. Taxonomists could form a pact in which they bound themselves over to classify consistently. Subjective classification would then have the stability of a naked pact, whose members agree to behave consistently; but the pact is guaranteed only by psychology or authoritarian decree, not by any argued principle.

In what I shall call objective classification, the choice of characters is dictated by a theoretical principle. The principle must specify some discoverable hierarchical property of nature, which it is desirable and technically possible for classification to represent. The principle should imply techniques of character choice, which define the kind of groups that the principle argues should be defined. What might such a principle be? We shall meet some strange possibilities in our historical chapter (Ch. 6); but in modern classification there are only two: phenetic and phylogenetic classification. There is a third principle, which I shall call teleological classification; but it has only been hinted at occasionally by biologists, it has not been developed into a full philosophy. Phenetic classification aims to represent a hierarchy of similarity of form among living things;

Evolution and Classification

phylogenetic classification aims to represent the branching hierarchy of evolution. Let us observe the two principles in action.

The evolution among a butterfly, a moth, and a wasp, was probably on the whole divergent and of approximately constant rate (Figure 1.1a). In this case the two kinds of classification are the same: the moth and butterfly are phenetically more similar (Figure 1.1b) and phylogenetically share a more recent common ancestor (Figure 1.1c) than either do with the wasp. But phenetic and phylogenetic classifications may also differ, for either of two reasons: convergence

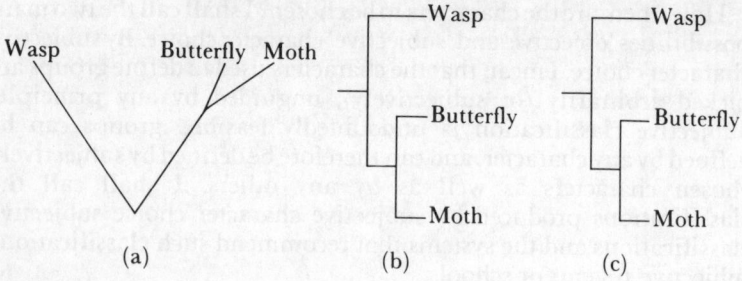

Figure 1.1
(a) The pattern of evolution of a wasp, butterfly, and moth. Their (b) phenetic and (c) phylogenetic classification.

and differential rates of divergence. Let us consider convergence first. Among a limpet, a barnacle, and a crab, the similarity of form of the adult barnacle and the limpet is convergent (Figure 1.2a). The limpet and the barnacle look more like each other and are classified together on the phenetic principle (Figure 1.2b); but the barnacle and the crab share a more recent common ancestor and are classified together on the phylogenetic principle (Figure 1.2c). (This example might be challenged. Barnacles and limpets only phenetically resemble each

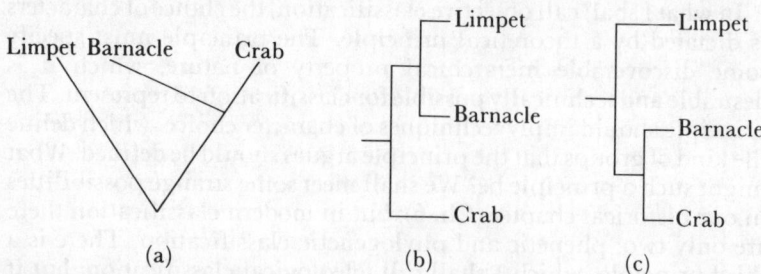

Figure 1.2
(a) The pattern of evolution among a limpet, barnacle, and crab. Their (b) phenetic and (c) phylogenetic classification.

The argument

other superficially and at the adult stage. A more thorough phenetic study might agree with the phylogenetic classification. Or it might not: a high degree of phenetic similarity is suggested by the fact that barnacles were classified as molluscs for many centuries until Thompson (1830, 1835) discovered their nauplius larva and 'demonstrated the class of animals to which they indisputably belong'. At all events, I hope the example may stand to explain, if not to illustrate, the principle.) A bird, a lizard, and a crocodile uncontroversially illustrate the conflict of the two principles produced by differential rates of divergent evolution. The birds share a more recent common ancestor with the crocodiles; but they have since undergone such rapid evolution that the crocodiles have been left looking more like lizards (Figure 1.3a). Crocodiles are phenetically closer to lizards: they both have four walking legs, scales instead of feathers, and are therefore classified together on the phenetic principle (Figure 1.3b); the phylogenetic principle classifies birds with crocodiles (Figure 1.3c). The two principles may agree or disagree, according to the pattern of evolution among the groups.

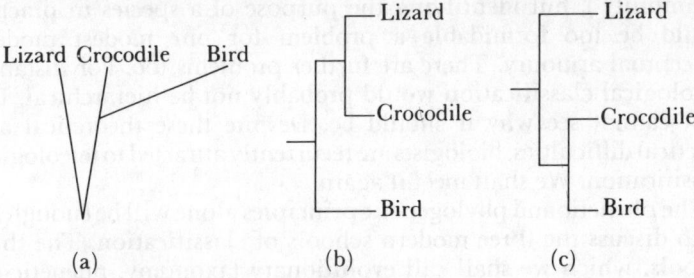

Figure 1.3
(a) The pattern of evolution among a bird, a crocodile, and a lizard. Their (b) phenetic and (c) phylogenetic classification.

Teleological classification cannot be so easily defined; but it is important enough to merit an effort. By teleological, I mean purposive; teleological classification seeks to group species according to their purpose in life, which in modern Darwinian terms means the function they are adapted to perform. Groups indicate not shared ancestry, nor shared simple similarity, but shared adaptation. A human analogy may help. We could classify human fabrications phenetically; but it might also be possible to identify the purpose for which each thing was made, and use that instead to define the groups. Take, for example, a pen, a word processor, and a television. The phenetic classification would, I imagine, group the television and the word processor; but a teleological classification might group the pen and the word processor, for both were invented for the purpose of

Evolution and Classification

writing. But it might not: it might group the word processor and television, because they have a shared purpose of making profits for electrical companies. And, indeed, practical ambiguity is the main difficulty with the teleological principle. Even if human fabrications do have unambiguous purposes, it is difficult to identify them. But be that as it may, a technological classification is one that aims to classify groups according to shared functions.

If the principle is difficult to apply to human fabrications, it is even more difficult to apply to living things. The purposes of living species are even less obvious than those of human fabrications. The word 'purpose' does, of course, mean something different in the two cases. Human fabrications are consciously planned before they are made; living species are not. But the term is not inappropriate, because natural selection does give living things a purpose, in the sense that they are adapted to perform certain functions. Species do have purposes, even if only as the sum of the adaptations of their members, and the ecological division of resources may even guarantee that each species has a unique, unambiguous purpose (or niche in Elton's (1927) original sense of 'the status of an animal in its community'), but identifying the purpose of a species in practice would be too formidable a problem for our modest modern conceptual armoury. There are further problems too. For instance, teleological classification would probably not be hierarchical. I at least cannot see why it should be. Despite these theoretical and practical difficulties, biologists are recurrently attracted to teleological classification. We shall meet it again.

The phenetic and phylogenetic principles alone will be enough for us to discuss the three modern schools of classification. The three schools, which we shall call evolutionary taxonomy, phenetic (or numerical) taxonomy, and cladism, favour different classifications in the three cases of Figures 1.1-1.3. Phenetic taxonomy and cladism are pure extremes; they prefer the phenetic and phylogenetic classifications respectively of the figures. Evolutionary classification, however, mixes the principles. It does not allow convergent groups and therefore prefers the phylogenetic classification (Figure 1.2c) of the barnacle, crab, and limpet; but when one group, such as birds, has evolved rapidly it prefers to classify it separately. Evolutionary taxonomy therefore prefers the phenetic classification (Figure 1.3b) of birds, crocodiles, and lizards.

Such are the meanings of the principles. But how, as principles, do they work? How would they enable objective classification? A principle must, for a start, be practical. In the examples of Figures 1.1-1.3, we simply assumed that the true phylogenetic and phenetic relations were known. If the relations were not discoverable they could not be used in classification even if it could be shown (by reason) that they were real. For the phylogenetic principle, some characters must demonstrably be good indicators of shared ancestry: those are the characters that should be chosen in classification. The phenetic principle must overcome the problem of character conflict, and classify by some measurement of 'true' phenetic resemblance; it might perhaps aggregate or average over large numbers of characters, or choose characters that are particularly good indicators of true phenetic similarity. A principle must also be unambiguous. Suppose we wish to classify three species, called *a*, *b*, and *c*. An unambiguous principle is one that allows only a single arrangement of the three species. An ambiguous principle would be consistent with more than one arrangement; it might allow $(a,b) c$, or $(a,c) b$, or $(b,c) a$, according to how the relationship was measured. If phenetic similarity could be measured in more than one way, for instance, the phenetic principle would be ambiguous. A principle can only be unambiguous if it specifies a real, unique relationship in nature. The phylogenetic principle (its justification will argue) dictates unambiguous groupings because only one true phylogeny connects all species. The justification of a principle must therefore show that the classificatory hierarchy corresponds to a unique hierarchical pattern in nature. If a school aims to define groups that are real, unambiguous in content, and technically discoverable, it may reasonably call itself objective.

A complete justification might ideally go further. It should show not only that its principle identifies a real and unambiguous property of nature, but also that the other principles do not. It will then explain why its own kind of characters should be used to define groups, and why those of the other schools should not be. A fully objective system should be the only possible principled system, for if there are two equally realistic and technically feasible systems, the choice among them would be subjective. It may turn out not to be possible to reduce the number of objective schools to one; but that is only an aspiration, not a necessity, for we shall have made some philosophical progress if we can show that some schools can reasonably claim to be objective.

In fact, all three schools would claim to be objective. They would also all agree that objectivity is desirable. As the desire is common ground, we need not spend time arguing for it, and from now on it is always a premise of my argument that objective classification is, other things being equal, superior to subjective. To expose subjectivity in a system is, henceforth, to criticize it. We can discuss the relative merits of the schools by considering whether the claimed objectivity of each school really is so.

I prefer the distinction of objective and subjective classification to another distinction often used in this kind of discussion: that of natural and artificial classification. The differing attempts of the three schools at self-justification all employ it; and we shall have need of it too. We must therefore fix its meaning. Its meanings are numerous indeed – they have been accumulating since Plato – and it would be disproportionate to discuss them all here (Ghiselin 1969, pp. 79–89; Kottler 1978; Panchen 1982 provides interesting remarks). For our purposes one simple distinction will do (Crowson 1970, p. 21). I intend it tentatively, because, as we shall later see, it can only be used rather informally. We must first distinguish between those characters of organisms used to classify them in groups, and those that are not. If (and I mean if), for example, mammals were defined by lactation, then lactation would be the character used to classify an animal, such as a baboon, as a mammal; all its other characters – its eyes, legs, heart, and so on – would be (at this taxonomic level) non-defining. An artificial classification is then one in which the members of a group resemble each other only for the characters that define the group; for other characters the members are uncorrelated. In the groups of a natural classification, the members not only resemble each other for the defining characters, but also for their other, nondefining ones; the defining characters are chosen to be representative of the rest. Taxonomists that study only one or two characters of a group will often classify artificially: you need to know many characters to classify naturally.

Stability is the practical advantage of natural classification. Many characters fall into the natural groups, and the groups should therefore not need to be changed every time the taxonomist studies some new characters. The more natural a classification is, the more stable it should be. Stability is an obvious virtue, for (other things being equal, cf. Gaffney 1979) classificatory changes are a biological nuisance. Moreover, all schools agree that natural classification is desirable; and each one even tries to justify itself by declaring that it alone, of all schools, classifies naturally. The controversy among the schools, therefore, might be decided by discovering which of them is truly natural, and several authors have suggested that stability in the face of new evidence is the best test of a classification (for example, Warburton 1967, and Mayr 1974 and McNeill 1983 boil down to the same position; if Wilson's (1965, 1967) 'consistency test' (cf. Colless 1966, 1967) were applied to a classification rather than a phylogeny it could be added to the list). Unfortunately the test is impossible. The concept is not powerful enough, for reasons we shall discuss in Chapter 3. All three schools can equally claim preference by this criterion, because, it will turn out, a single 'most' natural classification cannot be specified. Moreover, different taxonomists do not agree what the term 'natural' means. I have picked a phenetic meaning, and shall use it consistently; but it is only one meaning among several. Simpson (1961, p. 57), for instance, used a very different one, as he called a classification natural if it represented the process by which the various forms being classified had originated. Thus:

The argument

In spite of doubtful cases and myriads of complications, it is quite obvious to a modern scientist, as it was to a prehistoric Guariní Indian, that natural species do exist. The theoretical problem, then, is why they exist, and the solution of that problem must provide the criterion for whether a taxon does in fact correspond with a natural group. The general solution is now known ... Species exist because they evolved. That, in briefest form, is the natural reason for the existence of species and is therefore also the truly natural basis for classification.

When a word is used with such different meanings, we may be tempted to dispense with it altogether (Young 1983). But I shall follow a less radical procedure. The concept of natural classification may be too weak to resolve the taxonomic controversy, but it does merit discussion; I shall ask, of each school, whether it produces natural classifications in the sense I have specified.

So there is much more to classification than the definition of groups by characters. The disagreement of characters and the desire for objectivity (or naturalness), combined with the plurality of aspiring principles of objective character choice, have forced each system of classification, in order to maintain itself in competition with the others, to try to justify itself. Each of the three main schools of classification contains a justified principle of character choice, together with a set of techniques to recognize characters of the justified kind. It is the main work of this book to compare them.

That, however, is not the full limit of our subject. The last few years have seen the birth of what amounts to a fourth school, which I shall refer to as transformed cladism. It has stimulated various controversies both within biology and in more popular settings, but our concern will be with its techniques and justification. The former are easily identified, for they are the same as in traditional cladism. Its justification, however, is less clear. It is only clear what it is not. The original form of cladism (as we shall see) justified its techniques by the reality of phylogenetic relations. Transformed cladism, however, emphatically rejects the use of evolution in classification. The cladistic and transformed cladistic philosophies are exact opposites, and it is important to distinguish them. Although the two forms of cladism are often confused, we shall hold them apart terminologically by calling the traditional evolutionary form Hennigian cladism and the new form transformed cladism. Hennigian cladism deploys an evolutionary justification; transformed cladism strives for some non-evolutionary justification of the same cladistic techniques. Whether it can find one is a subject for later discussion. We need only now remark that transformed cladism, by its unique combination of techniques and philosophy, amounts to a distinct school; and that,

Evolution and Classification

because it proposes to separate evolution from classification, which two subjects have been closely related ever since evolution was first accepted, it is highly controversial.

But these unsubtle distinctions have not interrupted the noisy enthusiasm of Grub Street. There, what matters is the rejection of evolution. Transformed cladism, usually confounded with Hennigian cladism, has been greedily incorporated into that great media event, the crisis of Darwin. In that capacious melting-pot, the rejection of evolution from classification (Leith 1981, 1982) soon becomes the rejection of evolution (Davy 1981; Adley and Cary 1982; Hitching 1982). The similarly indiscriminate propaganda-hunters of the Institute for Creation Research have also discovered transformed cladism (Sutherland and Parker 1981), to provide, in turn, material for the Darwinian rabble-rouser Dr Halstead, who annexed it to the distributed typescript of his 1982 address to the British Association for the Advancement of Science. There is more still. Halstead, having sniffed (of all things) Marxism, successfully sparked a controversial correspondence in *New Scientist* and *Nature*, which in the latter burned hot and cold through 1979, 1980, and 1981, fanned at intervals by the *terribilità* of the editorial page.[1] The arguments of that public debate, however, have been too far removed from the central questions of classification for them to be of any interest to us here. We shall not need to refer to them again. We shall spend eight chapters examining the transformation of cladism, but the crucial arguments will all be taken from less popular sources.

The reader of the modern polemical literature may miss discussion of two subjects. I have little to say about either the status of palaeontology or the philosophy of Sir Karl Popper. Popperian philosophy is not a serious omission. It has been used mainly for rhetorical purposes, and Popper (1980) himself has remarked that the evolutionary disputes raging over his name have little to do with his philosophy (see also Cartmill 1981 and Hull 1983). We can safely omit them. The omission of palaeontology may be more serious. We shall discuss the use of palaeontology in cladism (in Ch. 4) and the discovery of ancestors (in Ch. 10) but not the disciplinary status of palaeontology. The controversial literature has considered the subject in depth. It has for a long time (we are told) been agreed that the purpose of palaeontology is to search for ancestors, and that the discovery of *Archaeopteryx* is the kind of act that palaeontologists are trying to emulate. Now, as the search for ancestors (we are assured) has been discovered to be unscientific, we are left with the question of

---

[1] The full run of correspondence and editorials can be tracked in *Nature* through the annual subject indexes, under 'cladism' and 'Darwinism'. Halstead's original article in *New Scientist* appeared on 17 July 1980; but then either the editor of that journal was less patient with his correspondents, or they less persevering.

The argument

what on earth, or in their museums, palaeontologists are to do. I do have opinions on this subject, but shall leave them implicit.

Even apart from the polemical literature, classification is a large subject. We are not going to cover it all. If we turn to any textbook of classification written before the modern controversy – such as *The Methods and Principles of Systematic Zoology* by Mayr, Linsley, and Usinger (1953, new edition Mayr 1969), Simpson's (1961) *Principles of Animal Taxonomy,* Crowson's (1970) *Classification and Biology,* or Sneath and Sokal's (1973) *Numerical Taxonomy* – we find, very properly, sections on such practical subjects as nomenclature, and even field techniques. These are essential parts of classification, but I shall say nothing about them here. This is a work on ideas, not practice. I am an evolutionary biologist, and my interest here in classification is confined to its relation with the theory of evolution. However, I have also omitted any discussion of what is probably the main theoretical issue of the earlier literature: the species concept. This, I think, marks a real difference between modern and earlier interests. The main development in post-Darwinian taxonomy was effected in the first two or three decades of this century and is often called the 'new systematics', after the volume edited by Huxley (1940). The new systematics was mainly concerned with the taxonomic levels below the level of species. The contents of Huxley's book reveal this, and it was explicitly remarked upon by Mayr (1942, p. 7) at the time, and others since (Blackwelder 1962; Sokal and Sneath 1963, pp. 5–9). The classification of the more inclusive categories was then regarded as a secondary problem. I think we can see why. Not only was the species their main interest, but the theory of the species that they developed, which they called the biological species concept, gave the species a special status in the taxonomic hierarchy. It explained the existence of (and defined) species by interbreeding, not similarity of form. If species were in practice recognized by their form, that was only because it indicates the theoretically important variable, interbreeding (Hull 1965). This definition of species decreases the interest of other taxonomic levels. Species may exist because of interbreeding, but no other level can. Other levels would have to be defined by similarity of form (the criterion rejected for species) and become relatively artificial concepts. The evolutionary biologists of the new systematics indeed almost universally agreed that the species was a relatively real category, and higher levels relatively artificial (Dobzhansky 1937; Huxley 1940, p. 3; Mayr 1942, pp. 280–91; cf. Simpson 1961, p. 57, who also disagreed with the biological species concept, on which see Mayr 1980).

In the modern literature, the relative interests in classification below and above the species levels have been exactly reversed. It is not that the discussion of the species concept in classification has ceased. The different schools have different species concepts – operational

Evolution and Classification

and phenetic in phenetic taxonomy, genealogical in Hennigian cladism – and they do reflect their different principles. But the modern literature is more concerned with the definition of the higher levels, where (as it happens) the different species concepts do not make much difference. It would be perfectly possible to substitute any species concept into any of the three or four taxonomic schools without making any difference to the debates that we are to consider. The arguments about species concepts are dispensable, and (in a short work) may be dispensed with.

Such is the nature of this work. I now wish to complete the introduction by running rapidly over the entire course of the argument, charting its main features, noticing its main conclusions, but not pausing to explain difficult points. I am going to assert arguments, not justify them, and I do not expect this chapter to be convincing; the real argument comes later. One purpose of the work is to decide whether transformed cladism is a coherent school, but I believe that question cannot be settled apart from the general controversy among all the taxonomic schools. Outside commentators and participants alike have too readily assumed that taxonomic schools are religious allegiances rather than philosophical decisions. They are not; and to demonstrate the fact we shall begin with a three-chapter survey of the three main schools, and of the arguments by which one school might be preferred to another. Philosophical justifications are an essential part of any taxonomic school, and I intend to draw attention to what they are, to stress their importance, and to place them in the centre of debate, in a position where they cannot be ignored.

We shall take evolutionary taxonomy first (Ch. 2). It classifies groups according to the course of evolution; its justification is that the process of evolution produces the natural groups of the natural classification. The correlated similarities of natural groups are due to common ancestry; but common ancestry cannot be observed; it can only be inferred, and it is inferred by the sharing of characters. Unfortunately groups may share characters not only because of common ancestry but also because of convergence. Evolutionary taxonomists have to distinguish between the two. They have to recognize homologies, which are ancestral characters, and exclude analogies (convergent characters) from classification. Their main techniques are to see, from other classifications, which characters do not appear to be liable to convergence, and functional analysis. The techniques of evolutionary taxonomy, however, are imperfect, and can be criticized as such. A further difficulty arises when it comes to translate the homologies into a classification. Evolution has two main aspects – the pattern of splitting and the pattern of phenotypic divergence. The two can conflict (Figure 1.3). Evolutionary taxonomists do recognize the difficulty, and they have discussed it. When the two

The argument

conflict, they prefer to classify according to phenotypic divergence rather than the order of splitting.

Numerical phenetic taxonomy (Ch. 3) was a reaction against the informality of the evolutionary taxonomist's techniques. It set out to eliminate subjectivity from all stages of classification. The doctrine that all characters are equal would replace the vague distinction of homology from analogy; the hierarchic output of multivariate statistics would replace classifications unstably based and subjectively defined by single characters; phenotypic similarity would replace the hierarchy of evolution. Such was the programme. Its justification was its formal objectivity. The system, however, turned out to be less objective than its proponents had believed. The subjectivity of the system was known to many critics, but was exposed with particular power by Johnson (1970). Objective classification must aim to represent a unique and real hierarchy; the aim, in the case of phenetic classification, must be a unique hierarchy of phenetic similarity. The difficulty is that, although any one numerical taxonomist can indeed produce a classification with any one multivariate cluster statistic, there are many different statistics (and an infinity of possible one) which do not generally give the same classification. Which statistic should be used? If the whole system is indeed objective then one of them should be the best, and that one should be used. At least it should be possible to show that in principle there is only one optimal cluster statistic, even if it has not yet been invented. But this is just what it is impossible to show. The reason is that there is no real hierarchy in nature of phenotypic resemblances: there are patterns of resemblances, some of them indeed hierarchical patterns, but there are an infinity of different hierarchies, depending on how you look at nature. No one is any more objective than another. Phenetic classification could only be made repeatable by a pact. It is subjective. The claimed advantage of numerical phenetics, its objectivity, is exactly what it lacks. It covers an infinity of repeatable techniques, but the choice among them is inevitably subjective; inevitable because the pheneticist seeks to represent a hierarchy that has no objective existence in nature.

The same criticism damages evolutionary taxonomy. That school too had sought to represent one aspect of phenotypic similarity. It differed from purely phenetic taxonomy in that phenetic taxonomy sought an aggregate measure of phenotypic similarity, whereas evolutionary taxonomy excluded similarity due to convergence and only sought to represent differential divergence; but the objection is the same. There can be no objective measure of differential divergence. It could be measured in an infinity of ways, and there is no hope of ever discovering a single natural measure of it. If groups are defined by phenotypic divergence, their definition is inevitably subjective.

Evolution and Classification

Chapter 4 examines the third school, cladism, in the original formulation of Hennig (1966), together with such later modifications as improve his system and remain within his philosophy. Hennig began with the question of what property of nature a classificatory hierarchy might represent. He reasoned that the pattern of phylogenetic splitting is the only known one. Classification should therefore represent it. His system soundly based, he went on to consider techniques by which the branching hierarchy might be discovered. He was led to the hierarchy of shared derived characters, which meant, in turn, that he needed techniques to distinguish ancestral from derived character states. Such techniques had been used, more or less vaguely, by evolutionary taxonomists, but Hennig and his followers have formalized them. In its justification, Hennigian cladism avoids the defects of the other schools, both of which are hung up on phenotypic similarity. By removing phenotypic similarity, Hennig made classification into a hierarchy of recency of common ancestry. The conclusion, at this stage, will be that Hennig's is the only known objective system of classification.

The remainder of the work considers some of the developments within cladism since Hennig. Some of Hennig's followers have not appreciated the importance of his philosophy. As we have seen, they have founded a new school, transformed cladism, and separated the cladistic techniques from their justification in phylogeny. They have tried to substitute for phylogeny a philosophy of 'information content', or Popperism, or even no philosophy at all. All of these changes make nonsense of the whole system. Chapters 5, 6, 7, 9, 10, and 11 oppose and destroy them. The various substitutes for the Hennigian philosophy can be understood as successive stages away from his phylogenetic system, but one slogan covers them all: the distinction of 'pattern' from 'process'. 'Pattern' means the distribution of forms in nature; the 'process' is the cause of the pattern. The 'pattern' is really the finished classification; the 'process' is really evolution by natural selection. Evolution, or phylogeny, is thus one theory of process (natural selection is another, in another sphere). When transformed cladists talk about the pattern – process distinction, they insist that the classification of 'pattern' must be kept separate from the study of 'process'. In similar style, they assert that it is perfectly possible to classify species without any help from any theories; which means (when we call a spade a spade) that the theory of evolution must be kept out of classification. After we have been through the progressive transformations of Hennigian cladism, we shall return to consider this pattern – process distinction (Ch. 11), only to find that it is invalid.

And what are the progressive transformations? Quite close to Hennig's system is the doctrine (Ch. 10) that cladistic classifications,

The argument

which are called cladograms, must be dichotomous. As we shall see, the only possible justification for this doctrine is that speciation is always dichotomous. But that kind of argument relies on a theory of process, a theory of speciation in fact, and is therefore unacceptable in transformed cladism. It tries to justify the dichotomy of cladograms instead as a 'methodological' preference, and by a higher 'information content' of dichotomous cladograms: these two arguments, it will turn out, are easy to destroy. A stage further from Hennig is the doctrine (Ch. 9) that the search for ancestors is unscientific. Ancestors indeed cannot be recognized by the techniques of cladism, if comprehended narrowly; and ancestors cannot be unambiguously represented in cladograms. But (it will turn out) the stronger assertion, that ancestors are an unscientific concept, is only cladism mistakenly hypostatized as science.

Chapters 5–7 examine the justification of transformed cladism itself. They need no space for its techniques, because its techniques have been carried over from cladism: the discussion in Chapter 4 will serve for transformed cladism too. The justification is less easily identified. We shall consider four possibilities. One is that it does not have one. Its advocates may first have realized that it is not necessary to assume evolution in order to apply the techniques of cladism; practical dispensability then led to philosophical rejection, and transformed cladism simply became Hennig's techniques without the justification that led Hennig to develop them. But the rejection of Hennig's justification is not itself a justification. If transformed cladism simply cut away Hennig's techniques from the reason why those techniques (rather than any others) should be used, it would be unjustified. It would be nakedly subjective. The whole school would be reduced to a peculiar faction of phenetic taxonomy, with its own bizarre cluster statistic, operated for no more reason than that it has agreed to use it. The choice is, without Hennig's philosophy, utterly arbitrary: there is no reason why anyone else should use the cladistic cluster statistic rather than the dozens of others that are available. It is awkward to use and might be rejected on grounds of impracticality alone.

But transformed cladism is not a single articulated system. While my initial reaction was that transformed cladism did not even unconsciously realize the need for a justification, and could as such be rejected, closer inspection suggested that it had offered three different self-justifications. The first is a philosophy of 'absent' characters, which we shall deal with in Chapter 5. The second (Ch. 6) is that evolution need not be assumed in classification because it had been successfully performed much as it is today for two millennia before evolution was an assumption. The third (Ch. 7) is that evolution must not be assumed in classification, in order that classification may itself be used to test evolution; transformed cladism, in other words, is needed for the important purpose of providing a non-circular test of evolution.

The justification by absent characters is a simple verbal trick. Reduced to its simplest form, it runs like this. Cladistic techniques distinguish ancestral from derived character states, and only the latter are used to define groups in the cladogram. Derived characters are the only characters in the classification. Now, non-cladistic groups are by definition not defined by derived characters, because if they had been they would be cladistic groups. The verbal trick is to miss out the word 'derived' from the last sentence; it then says that non-cladistic groups are not defined by characters, and we are apparently forced to use cladistic techniques in order to define groups at all. Reduced to this form, the trick is obvious, but to disentangle it from the published literature will be a more laborious task.

The mere fact of the historical conduct of classification without the theory of evolution shows only that it is practically possible (which is not in doubt), not that it is objective (which is). In order for history to justify transformed cladism, it must fulfil two criteria. The techniques of pre-Darwinian classification must have been cladistic; and their operators must have possessed a valid justification of their practice. Its validity, moreover, must be relative to 1980, not 1800. We shall examine the classifications of several important pre-Darwinians – Linnaeus, Cuvier, Owen, von Baer, Agassiz – to see if they meet the criteria. They meet neither. Their techniques are closest to those of evolutionary taxonomy; and their justifications are either essentialist or theological. The chapter also has a positive purpose. It will illustrate (in the theological justification) how a justificatory principle must be not only realistic but also practical; and will consider (in the essentialist justification) a principle other than the phylogenetic and phenetic ones we have otherwise treated as the only alternatives.

The third justification proposes that transformed cladism is a test of the theory of evolution. We shall tackle it in four stages. They are to ask: (1) Whether classification can test evolution. (2) Whether classification is needed to test evolution. (3) Whether the required classification must be cladistic. (4) Whether it is circular both to assume evolution in classification and to use classification to test evolution. If transformed cladism is to be successfully justified as a test of evolution, all four questions must be answered in the affirmative. But in fact only question 1 will be: questions 2, 3 and 4 will all be answered in the negative. Transformed cladism, therefore, cannot be so justified.

With transformed cladism out of the way, it only remains for us to complete the work of destruction on its associated philosophy, the distinction of pattern from process. Evolution is a theory of process, and is needed in classification. But what of the process that causes

The argument

evolution, natural selection: is that needed in classification? The theory of natural selection is a part of functional analysis, which (as Ch. 2 will reveal) is used in evolutionary taxonomy to distinguish homologies from analogies. Analogies are caused by convergence, and convergence is caused by natural selection; if therefore we can recognize, by functional analysis, which characters are likely to be convergent, we can (by elimination) recognize probable homologies. There is what might be called a functional criterion of homology (the name being by analogy with the other criteria of homology, such as the embryological criterion). It might also provide a technique of distinguishing ancestral from derived character states. It would have to be shown that, for a pair of character states a and a', natural selection could drive evolution from (say) a to a', but not vice versa; a' would then probably be the derived state. Many cladists have denied arguments of this kind. If they are correct to do so, one theory of process at least, natural selection, would not be needed in classification. Chapter 8, however, will argue that natural selection should not be so dismissed: it can supply a valid functional cladistic criterion. It can provide one of several techniques (besides those covered in Chapter 4) by which ancestral and derived character states may be distinguished. None of them is perfect, and the functional criterion is only one criterion among several. It is not necessary to use it, because other techniques do exist. But it is not necessary to use any of them. The question is not whether it is necessary, but whether it is sensible to use it; and the answer is affirmative. Because it is not easy to distinguish ancestral from derived character states, we should use all the lines of evidence at our disposal. To exclude the functional criterion simply because it flouts the ideology of pattern and process would be self-destructively doctrinaire.

In summary (Ch. |11), we shall look directly at the distinction of pattern and process. What is the proper relation of the theory of evolution and the classification of life? The best known system is Hennig's, enriched by all subsequent developments that improve its techniques but do not ignore its justification. In this system, the theories of evolution and of natural selection are a crucial part of the method by which we search for pattern: evolution provides the system's justification, forms part of its technical logic, is the fountain of new ideas, and of refinements of old ones.

As cladists have tried to separate classification and evolution, abandoning ancestors, providing strange justifications for dichotomous cladograms, straining at the functional criterion, and entirely removing evolution from their philosophy, they have insensibly engaged themselves in the same snares of positivist, puritanical absurdity as captured numerical phenetics two decades before. Numerical phenetics too had tried to remove the theory from classification, to leave it a pure, rigorously empirical search for

Evolution and Classification

pattern. Most taxonomists pass their lives in practical research, and worry little about its theoretical setting. But from time to time some sage looks up from his museum work-bench and observes, in the light above, the parasitical, unnecessary, cancerous growths of theory which (as he imagines) are supported on the labour of taxonomy. He determines that the cancer must be cut out; and taxonomy restored to health and power. He communicates his resolve, and, if his colleagues are receptive, one of those great, puritanical revolts will swell up from the work-bench and sweep through the museum. From then on, the course of history is fixed. Demands will be issued that the parasites be expunged, and power returned to its true source. Just listen to the conclusion of Platnick's (1979) revolutionary pamphlet:

> So what Hennig may well have done in general (and may perhaps have even set out to do) is to demonstrate the inadequacy of the syntheticist paradigm, by showing us that we are hardly likely to achieve any understanding of the evolutionary process until we have achieved an understanding of the patterns produced by that process, and that even today we have hardly begun to understand those patterns.

The progress of the revolution will then depend on whether the biological community has been caught off guard. In this respect, numerical phenetics had the advantage. That enthusiasm had swept through dozens of museums and even penetrated a few universities before a reaction came. Puritanical systems of classification, however, are always vulnerable to internal fragmentation and external philosophical reaction; both of which, in the end, are inevitable. Some Menenius Agrippa will stand up and persuade the rebels that, just as limbs have need of their belly, so too does science have need of its theories. Good science is ridden with theories and flourishes on imaginative hypotheses: and just as numerical taxonomy in time yielded to the attentions of philosophical critics (Ghiselin 1966; Hull 1967, 1968; Johnson 1970), so too will transformed cladism. Transformed cladists may, like all revolutionaries, declare that they are returning to an earlier period of liberty, in this case a state of pre-Darwinian innocence. But much water has flowed under the bridge since then: pre-Darwinian classification has been lost down the stream of history; it cannot be recovered. Arguments, lethal to that system, have been put into print since then. I would make little claim to originality as I fetch a philosophy from an earlier literature, and apply it to a new situation.

2
The techniques and justification of evolutionary taxonomy

Evolutionary taxonomy has been developed by the best philosophers of classification and has dominated the activities of the museum work-bench for over a century. It possesses both a reasoned philosophy and practical techniques; it has accumulated a rich and broad, tolerant and flexible set of concepts, and a huge literature. We need concentrate, however, only on its leading ideas, which can be found in such standard works as *The Methods and Principles of Systematic Zoology*, by Mayr, Linsley, and Usinger (1953, new edition Mayr 1969), and *Principles of Animal Taxonomy* by Simpson (1961).

The justification of evolutionary taxonomy is simply expressed. Evolution has produced the natural groups of life (Sprague 1940); classification should seek to be natural rather than artificial; classification should therefore represent evolution. Evolution results in natural groups because all evolutionary change takes place in the same phylogenetic tree: all changes in all characters must have taken place within the same pattern of lineages. The distribution of character states will define a pattern of groups according to where in the phylogeny they arose. If a character defines a monophyletic group (that is, one with a common ancestor that also possessed the character), other characters, which changed at the same time, should also fall into the same grouping. Other characters may be distributed more narrowly within it (if they changed later) or more widely (if they changed earlier), but, except for convergence, they should not contradict it: they should not be found in some members of the group, and in some members of another group. If, however, different characters were not forced to change within the same phylogenetic tree, taxonomic groups would not be natural. If, in some scarcely conceivable non-evolutionary system, different characters changed within different patterns of lineages (joining the same set of species in the end), they would not end up distributed in the same groups. The case of convergence illustrates the point. Convergent characters do

Evolution and Classification

not define monophyletic groups; they define polyphyletic groups, whose common ancestor did not possess the character. There is no reason why other characters should (except for functional correlation) fall in the same groups: groups defined by convergent characters are liable to be artificial.

If the groups of the classification are to be evolutionary, the course of evolution must be known. Its discovery is the technical problem of evolutionary taxonomy. The material available consists of such characters of species as may be observed in the museum, or (exceptionally) in the field. What can the evolutionary taxonomy do with these? It might just assume that shared characters between species indicate shared common ancestry; and the greater the number of shared characters, the more recent the common ancestor. But that simple rule is not enough. Because some characters are convergent, different characters often give conflicting indications. If convergence is ignored, polyphyletic groups may be defined, which, as we have seen, are unacceptable in evolutionary classification. Much of the practical work of evolutionary taxonomists has been to remove polyphyletic groups from earlier classifications. A typical example of convergence is that between the marsupials in South America and Australia, and the placentals elsewhere. In this case, an Argentinian *autodidáctico* in 1906 had exaggerated the age, and overlooked the convergence, of his native fossil fauna, and concluded that almost all the mammalian groups had originated in his homeland (Simpson 1980, p. 14). That is no longer believed; but it took several evolutionary taxonomists several years of work to remove the convergent groups from mammalian classification.

Once it is accepted that convergence must be kept out of classification, the whole enterprise of evolutionary classification becomes less easy. Taxonomists cannot use any characters to define groups; they may use only non-convergent characters. And if convergent characters are to be spurned they must first be recognized, for convergence is too common a phenomenon to ignore. 'The first step, then, towards the achievement of a phylogenetic classification is an analysis of the taxonomic characters to determine which of them are derived from common ancestors (*homologies*) and which are spurious similarities (*analogies*), usually convergent adaptations correlated with similar habits' (Mayr, Linsley, & Usinger 1953, p. 42).

The principle of evolutionary taxonomy may be to exclude convergent characters (and polyphyletic groups); its practical realization is another matter. In their published accounts, the methods are not rigorous; they are often vaguely formulated, and without any pretence to the contrary. But even if they are vague, they do exist. We shall discuss them only incompletely in this chapter because the techniques used in evolutionary taxonomy have been improved by numerical taxonomy and cladism, and we shall

The techniques and justification of evolutionary taxonomy

elaborate the most modern techniques later. We do not need to know their primitive forms, but should give evolutionary taxonomy some credit for the techniques to be discussed in Chapters 3 and 4. For now, let us consider two kinds of technique to reveal convergent characters, by theory and by observation.

By theoretical analysis, I mean the application of the theory of evolution to estimate the probability that each character is homologous or analogous. Two kinds of criteria have been used. Both were discussed by Darwin in the *Origin* (1859 [1969 edn, pp. 399–403]). According to the first criterion, non-adaptive characters are more likely to be homologies than are adaptive characters; and according to the second, characters uniformly adapted to a broad range of environments are more likely to be homologies than are characters that fickly change, among a limited number of states, with every small change in the environment.

Let us take the first criterion first. Convergence is caused by the natural selection of adaptations; the obvious method of avoiding convergent characters, therefore, is to avoid adaptations. Groups should be defined by non-adaptive or, more precisely, selectively neutral characters. Whatever its merits, the idea has certainly been influential. Nearly every enthusiasm that has swept through evolutionary taxonomy has been inspired by the doctrine of selective neutrality. Developmental characters and molecular sequences are two outstanding instances.

In the late nineteenth century, a biologist would have classified his material by one (or both) of two main kinds of character – adult characters and developmental characters. Different biologists held different opinions about which of the two should be preferred, and the preferences changed with time. The golden age of the embryological criterion began in about 1870, and ran for more than two decades. Embryologists made hay in that long summer undisturbed by any murmuring sceptics. The doubters did come, of course, but they do not matter to us (see Gould 1977a; Ridley 1986). What we should ask is why biologists in 1870 held embryological evidence in such high esteem. Let us turn to the principal paper of its principal British advocate: the paper on the 'genealogical classification of animals', by Edwin Ray Lankester (1873). As Lankester observed the classifications of his time, he saw, in an updated form, the methods and the results of Cuvier; species were classified according to the comparative anatomy of adults. Lankester disapproved of what he saw. He preferred the method of embryology, and

> It would not be surprising if the facts of development were to lead to another primary grouping of the animal kingdom than that indicated in the four Cuvierian types or the six or seven types now generally adopted ... They are confessedly

groupings based upon the anatomy of the adult organism; and therefore necessarily there has been a tendency in forming them to attach great importance to distinct plans of structure due to a secondary adaptation, whilst the fundamental community of organization has been ignored with something like intention (pp. 323-4).

According to the 'recapitulation hypothesis', however, developmental characters are often ancestral, 'this tendency to recapitulate' being 'the fullest expression of the phenomenon termed heredity'. Recapitulationists such as Lankester were quite sophisticated when they came to discuss how natural selection controlled development, but we need not go into that here (see Ridley 1986). They preferred embryological evidence in classification because they thought adult characters unreliable, because they were adaptive and liable to convergence; developmental characters, however, were sometimes ancestral. No modern biologist would agree exactly with Lankester, but that is not the point: the nature, not the validity, of his argument is what matters. Developmental characters were to be trusted because they were, under certain circumstances, unaltered by natural selection.

In modern molecular taxonomy, the character is different, but the inspiration is the same. If the proteins, or DNA sequences, of species were thought to be adaptations then they would provide 'just another character' for use in classification. As such, they have the practical disadvantage, relative to other characters, that they are difficult to see, but the advantage that they are easy to compare. It is difficult to compare the differences between two limbs with the differences between two skulls; but it is easy to compare the differences of cytochromes with those of haemoglobins. The large number of molecules is also an advantage. These advantages and disadvantages, however, have little to do with the popularity of the technique. That is due to the 'molecular clock'. The molecular clock is a theory concerning the rate of protein evolution, according to which proteins evolve at an approximately constant rate. The difference between the proteins of species, therefore, should be exactly proportional to the recency of their common ancestry: species with more similar proteins should have a more recent common ancestor. I do not wish to discuss here whether the theory is true; the quantity of evidence is too great. Let us concentrate on the justification. Why should molecules evolve at a constant rate? The answer, of course, is that molecular evolution is selectively neutral (Kimura 1968, 1983). Molecules are good classificatory characters because their evolution is non-adaptive. (I am not denying that selective explanations of the molecular clock have been devised; but I would deny that they have inspired molecular systematics.)

The techniques and justification of evolutionary taxonomy

With the wisdom of hindsight, we know that Lankester was wrong (we shall return to this later as well, p. 67). So too was anyone else who thought that embryological characters were non-adaptive, and, for that reason, of especial classificatory value. It is too early to judge molecules. We cannot yet say whether classification by molecular evidence will be superior than had similar efforts been made with other kinds of evidence. However that may be, the argument itself has run a similar course, one century on, to the biogenetic law. If Fritz Müller's *Für Darwin* was published in 1863, and Ernst Haeckel's *Generelle Morphologie* in 1866, Kimura's important paper was in 1968; if the great age of biogenetic enthusiasm was in the 1870s and 1880s, we know the 1970s and 1980s; if the support for the biogenetic law was crumbling in the 1890s, well... As molecular taxonomy progresses, it will move on to another stage. We cannot predict what it will be, or when it will come; but if any molecular taxonomists dislike my historical analogy, I would offer them some comfort in the prospect of a long period of power. The theory of recapitulation was still influential in the 1920s, and even in the 1930s biologists could still win their spurs in combat, if a little uneven by now, with the shade of Haeckel.

Developmental and molecular characters are not the only characters whose taxonomic merits have been argued in this way. Classification by animal signals is another instance. According to a classical theory, animal signals are so distorted during their evolution that the final relation of the form and meaning of the signal is quite arbitrary. Therefore, as Lorenz (1941 [1971, p. 19]) argued in his great work on the classification of the Anatidae by their displays, signals (and 'releasers' in particular) should be good taxonomic characters:

> There is a second factor which makes releasers particularly useful in phylogenetic considerations: because the special form of a releaser is not directly derivable from its function (in contrast to mechanically operative motor patterns) and is not influenced by the latter, the *possibility of convergence* can be fairly confidently excluded when there is a correspondence between the elicitatory ceremonies of two related species.

Lorenz's idea, as it happens, never really caught on; but the nature of his argument is revealing. He knew how he had to justify a proposed new class of classificatory evidence. He had to show that natural selection would not cause it to converge.

The taxonomic interest of developmental, molecular, and behavioural characters followed, in all three cases, from their supposedly non-adaptive nature. That is historical fact. But the argument itself is not convincing. Its premise can always be challenged. It is scientifically difficult, perhaps even impossible, to prove that a character is selectively neutral, because neutrality is a negative or 'null' hypothesis. Null hypotheses can in principle be refuted – they are good hypotheses in the Popperian sense – but that is little help to the scientist.[1] Neutrality can be confirmed with respect to a particular environmental variable, or even a series of environmental variables; but the danger will always remain that someone will demonstrate that some other environmental variable is selectively influencing the character. The hypothesis of neutrality will then be refuted. This is as true of embryological and ethological evidence as of molecular. In the wake of the theory of recapitulation came doubts as to whether embryonic characters were any less adaptive than adult ones (Garstang 1922, 1928). Many students of animal signals would doubt whether the relation of form and meaning in signals is arbitrary (e.g. Morton 1977; Hamilton and Zuk 1982). It is at present controversial whether molecular evolution is controlled by selection (Goodman 1981). Many characters that have been thought to be non-adaptive have turned out, on investigation, to be adaptive; and judgements of non-adaptiveness are precarious generally (Cain 1964). Modern evolutionary taxonomy may prefer not to justify its techniques with such a shaky philosophy. It may prefer to rely instead on the second reason why some characters might be more conservative in evolution than others.

It is that some characters are adapted to a wider range of environments than are others. Broad and generally adapted characters will not change as much in evolution as will narrower, more specialized characters. The more slowly a character evolves, the more it reveals its ancestry, and the better a taxonomic character it provides. This kind of evolutionary constancy does not result from non-adaptation: it depends instead, as Darwin (1859 [1969 edn, p. 400]) said, 'on such organs having generally been subjected to less change in the adaptation of the species to their conditions of life'.

Broad adaptation is a better justification of the taxonomic value of a character than is selective neutrality. It is theoretically more plausible. Many modern biologists are aware this is so, and, indeed, probably often intend it when they describe a character as non-adaptive: the translation is easy (Simpson 1961, pp. 100–1). But biologists do often treat ancestry and adaptation as alternatives. We could just let this pass; but it may be worthwhile to point out one of the reasons for the mistake. Most evidence, and especially crude evidence, of adaptation is provided by the comparative method. If one character repeatedly evolves in one particular environment, we

---

[1] Which has not stopped some Popperians from arguing that neutralism is unscientific (Ayala *et al.* 1974); Maynard Smith (1978, p. 37) points out that 'the neutral theory is a good "Popperian" one'.

properly infer that the character is an adaptation to that environment. We necessarily lack this kind of evidence for characters that do not often change in evolution: and these are the ancestral characters. We therefore usually cannot infer, by the comparative method, how ancestral characters are adapted. This mistake is to confuse the inapplicability of the method with the absence of the subject it investigates: one kind of evidence of adaptation is absent for constant characters; but it does not mean they are not adaptive. The gnawing apparatus of the Rodentia illustrates the point (Cain 1964, p. 26). It is constant in all rodents, but it is certainly adaptive; it is an adaptation for feeding. Methods other than comparison, such as functional morphology or experiment, can reveal how evolutionarily constant characters are adapted to their environments.

Although relative evolutionary conservatism is a theoretically sound method by which to distinguish taxonomically good characters from bad ones, it is difficult to apply in practice. It can be applied only rather loosely. No one doubts, for instance, that a backbone is a conservative character, and that plumage colours are not. But it is one thing to prove that the distinction is meaningful and theoretically valid; and quite another to provide a method by which all, or even many, particular characters can be quantitatively valued. The general direction of the rule can be stated – backbones are good characters, plumage colour bad – but that is not as complete a taxonomic technique as is desired.

So much for the theoretical method. Let us turn now to what I called the 'observational' one. It, too, has two main forms. The justification of both is that if a character has been proved to be reliable in one case, it will probably be good in another. The method assumes that reliability is a property of a character; therefore, if a character is reliable in one group, it will be in others too. This could be justified by the arguments about broad and narrow adaptations that we have just considered. If some characters do indeed fit a broad range of environments, and change less often, than others, then they are the reliable characters that evolutionary taxonomy seeks to recognize. It is, as we have seen, difficult to recognize them by theoretical analysis. But if a character is a broad adaptation it will remain constant in form, which should be detectable by simple observation. The argument of what I am calling the observational method is as follows. The constancy of a character among a set of species suggests it is a broad adaptation. If the breadth of an adaptation is more a property of a character than of a taxonomic group, then what is a broad adaptation in one group will probably also be one in other groups. If a character is particularly constant in existing groups (and is therefore likely to be homologous), it is also likely to be homologous in other groups. Such a character can accordingly be relied on in the classification of many groups. Of course, the group may matter: a

Evolution and Classification

character might be a broad adaptation in one group, but not in another. The method supposes this to be relatively unimportant.

The observational method, therefore, is to discover which characters are recommended by the study of other specimens, and to rely on them. If the characters are found in fossils, the method could theoretically be used even in the absence of any previous classification. Let us suppose that we have scored our specimens for their numbers of tail vertebrae, numbers of neck (cervical) vertebrae, tooth shape, and dental formula. All these characters are preserved in fossils. If some are more constant than others, if the number of neck vertebrae, for instance, was more constant than the number in the tail, that would be evidence they would be more reliable characters in other specimens. In practice the fossils will have been classified already, which assists the test of constancy because the test is strongest for a known sequence of ancestor-descendant populations (that is, the samples are classified in one taxon). The same method is possible without fossils. Instead of fossils, it can use the evidence of other, previously existing taxonomic groups. The fit of a new character to established taxonomic groups measures the character's reliability. If the character fits a group, being constant across the range of environments occupied by the species of that group, then it is probably reliable; if it changes within the group as if it had been repeatedly evolved and lost, it is probably unreliable. The strength of the test depends on how well established the existing groups are. If the groups are fairly certainly evolutionary, rather than convergent, and a new character fits them, that is strong evidence; if the group is newly erected, tentatively defined by only a few characters, and a new character fits it, that is less strong evidence. In practice, increasingly natural groups are developed, which provide increasingly strong tests of the reliability of new characters. The test simply checks whether a new character has the same taxonomic distribution as other characters, for the groups that the new characters are tested against are themselves defined by characters.

The coelom may provide an example. In the nineteenth century, the coelom was found to define a natural group of animals. The Polyzoa, however, in their original extension, comprised two groups, the Entoprocta and Ectoprocta. The former possess a true coleom but the latter do not; the Entoprocta are pseudocoelomate. For this reason above all, Hyman (1951, p. 551) insisted on their separation: 'the insuperable difficulty [for their inclusion in one phylum] is the pseudocoelomate nature of the body cavity of the Entoprocta whereas the Ectoprocta are typical coelomate animals'. The coelom was known to be a good character from other animals and therefore could be supposed to be here too. The separation favoured by Hyman, however, is not universally accepted (Nielsen 1971, 1977); and the coelom 'may be polyphyletic' (Clark 1964, p. 216).

The techniques and justification of evolutionary taxonomy

The method is not valid unless the character in question has been tested, whether on pre-existing groups or fossils. It must have been shown to correlate with other characters, to define a natural group. If it was used to define the pre-existing group, other characters must be shown to have the same taxonomic distribution; if it was not originally used to define the group, it must subsequently have been found to fit the group well. If a character has been so tested, and has passed the test, we have evidence that it is homologous. In the manner already described, we can then reason, provisionally and in proportion to the case, that it is probably homologous in other, unclassified specimens.

I say *provisionally*. Evolutionary classifications are always subject to improvement. Each new piece of evidence can be used to test the existing groups, to test the current ideas about which characters are homologies and which analogies. If evidence piles up against a classification, it can be changed; and the revised classification will be superior to the one it replaces. Of course, the taxonomist's initial crude ideas of the classification of a group will influence the assessment of whether the next character under study is taxonomically good or bad: errors may be perpetuated, 'but the effect of the early classification *decreases* as re-classification takes place in the light of new evidence' (Hull 1967, p. 180). The whole process is one of 'successive approximation'. 'Classification and re-classification ... goes on all the time in evolutionary taxonomy in the light of the discovery of previously unknown species and additional evidence' (Hull 1967, p. 179).

Critics of evolutionary classification – and especially the numerical taxonomists we shall meet in the next chapter – have often mistaken this procedure of successive approximation, or of 'reciprocal illumination' (Hennig 1966), for argument in a circle (see especially Hull 1967). The process clearly is not circular, but a normal scientific sequence of an initially crude hypothesis, a test, a hypothetical extrapolation, a further test, and so on. It is a process of theory-building. Its critics were led to their conclusion by a different understanding of evolutionary classification. Sokal and Sneath (1963, p. 7), for instance, remark that

> circular reasoning arises from the fact that new characters, instead of being evaluated on their own merits, are inevitably prejudiced by the prior erection of [another] taxon A based on other characters (X). Such a prejudgment ignores the fact that the existence of A as a natural (or 'monophyletic') group defined by character complex X has been *assumed* but *not demonstrated*.

In other words, according to Sokal and Sneath, characters are not tested before being used to assess new specimens. Although it sounds unlikely, untested characters may sometimes have been assumed to be homologous. But the misuse of a method is not a reasonable objection to it because any method, be it ever so good, can be misused. The passage just quoted 'is not a characterization of the best in evolutionary taxonomy but a parody of the worst' (Hull 1967, p. 178). Circular reasoning is only a possibility in evolutionary taxonomy, not an inevitability. The system should be philosophically judged in its best form. Characters can be tested against other characters, and even against unclassified evidence, such as from fossils. Once it has been shown to be homologous, that demonstration can be put to work. Any proposition in science, once proved, can be assumed in further investigations. Sokal and Sneath (1963, p. 22) could not see the difference between 'the principle of reciprocal illumination' and 'the much-condemned vertical construction of hypothesis upon hypothesis'. The difference is that between stacking a hypothesis on a successfully tested hypothesis and stacking a hypothesis on an untested hypothesis.

Opinions differ on the proper parts of the theoretical and observational techniques within evolutionary classification as a whole. I think of the observational method as testing the reasoning of the theoretical one. I therefore think they should be used together. Some authors, however, think the reasoning used in the theoretical method so imprecise as to be contemptible, and prefer to recognize homologies as those characters that are observed to have the same taxonomic distribution as a majority of other characters. They would first cluster the species according to all the characters, to give the most 'parsimonious' groups; only then, after the species are classified, would they recognize the homologies as the characters that fit the groups and the convergent characters as those that do not. This is a step towards the numerical taxonomy of the next chapter. In the extreme, all the characters could be fed in without prior analysis; but there is a gradation of methods from this extreme through to the individual testing of characters by the theoretical and observation techniques I have described. Other authors distrust the unanalysed statistical evidence of many characters, and prefer to rely on fewer, thoroughly thought-out characters. According to Hecht and Edwards (1977, p. 13; and Gaffney 1979, p. 99 agreed), for example, 'only one character is needed ... From the viewpoint of lineage detection, it is more important to use a few well-analysed morphoclines than many poorly ... analysed ones'. Philosophers differ, but taxonomic practice is fairly uniform. Faced with a conflict of characters, evolutionary taxonomists almost invariably try to understand which characters are more reliable. Parsons and Williams (1963), for instance, in an important paper on the relationships of the modern Amphibia, discuss a long list of characters, to try to understand whether the similarities among groups are convergent or ancestral. It would be laborious, and probably unnecessary, to prove that Parson and Williams are not exceptional; they illustrate a rule. Biologists trying to discover homologies think about characters; they do not just multiply them.

After the homologies have been recognized, the next stage is to classify the species into groups according to their homologies. Although this can be done easily enough, and often without ambiguity, some groups pose a conceptual paradox. The classification is intended to represent evolution: but evolution has more than one aspect, and its different aspects do not always imply the same classification. An evolutionary tree cannot be simply converted into a classification. As Darwin said of Charles Naudin, 'his simile of tree and classification is like mine (and others), but he cannot, I think, have reflected much on the subject, otherwise he would see that genealogy by itself does not give classification' (Darwin 1887, vol. 2, p. 247; see also Hull 1970). Darwin's remark is correct for a number of reasons of which only one matters here. Evolution has two main aspects: the order of splitting and the rate of divergence. As we saw in Chapter 1 (Figures 1.1-1.3), they can either agree or disagree. If splitting and divergence both take place at approximately constant rates, there is no difficulty. The paradox arises if divergence and splitting do not take place at constant rates (Figure 1.3). Bird and crocodile share a more recent common ancestor than do crocodile and lizard. Classification by order of splitting puts a crocodile with a bird; classification by degree of divergence separates the bird from the group of crocodile and lizard. When the different aspects of evolution disagree, evolutionary taxonomists have to decide which to represent. They have indeed decided. They have consistently chosen to classify groups according to their degree of divergence, not order of splitting. But how has the apparently ambiguous aim to represent evolution led to a consistent practice of representing one aspect of evolution rather than another?

At this stage, we shall merely set out the reasons given by evolutionary taxonomists. We shall not judge them until the end of Chapter 4, after we have examined the competing arguments of phenetic taxonomy and cladism. Within evolutionary taxonomy itself, we should distinguish the justifications offered before and after the rise of cladism. They differ not in their logic, but in their decisiveness. Originally, evolutionary taxonomists recognized a difficulty; and, after some inconclusive reasoning, in which they pointed out that phenetic divergence is a real component of evolution, they came down in favour of the classification based on phenetic divergence (Figure 1.3b) (Mayr, Linsley, and Usinger 1953, pp. 45, 168-77; Simpson 1961, pp. 129-32 and elsewhere in that chapter). If the differential rates of evolution are large enough, the evolutionary taxonomist would allow it to override the order of

Evolution and Classification

branching of the group. Birds are raised to high taxonomic rank, equal with that of all the reptiles, because of their rapid evolution into an aerial niche. The difficulty with this argument is that it applies as strongly to convergence as it does to differential divergence: for convergence is a real part of evolution too.

The aim can be re-expressed by means of two crucial terms: homology and monophyly. In these terms, evolutionary taxonomy aims to classify monophyletic groups, defined by homologies. There are, however, homologies and homologies, monophyly and monophyly. Some homologies are evolutionarily ancestral, and others derived. Within the group of birds, crocodiles, and lizards, feathers are homologous within birds and scales are homologous within crocodiles and lizards, but scales are the ancestral homology of the whole group, feathers are a derived homology within birds. The evolutionary taxonomist classifies according to homologies in general, using ancestral and derived ones indiscriminately. A case like that of the birds and reptiles arises when one group evolves many unique, derived homologies of its own, while the group from which it arose retains the homologies of its ancestors. The birds have evolved many unique homologies while the reptilian groups have retained many ancestral homologies. The reptiles accordingly possess many homologies, and so too do the birds. They are both good groups in evolutionary taxonomy, which takes no notice of the fact that they are defined by different kinds of homologies. I shall use the team homology like evolutionary taxonomists, to refer to both ancestral and derived characters. Some cladists (e.g. Patterson 1982b) would restrict the term 'homology' to derived homologous similarities; but I shall not follow them in this work because the broader meaning is coherent in theory and, I believe, more widely comprehended. There is not much to be gained from, and little hope of effecting, the change.

The evolutionary taxonomist defines groups by homologies in order to ensure that the groups of the classification are monophyletic. This term is also ambiguous. Fortunately we do not need to discuss all its proposed meanings (to which Holmes 1980, pp. 55-73, gives many references). For our purposes we need only distinguish two important meanings, although I shall mention two others as well. One unimportant meaning can be swiftly dismissed. It will not do to 'employ the term monophyletic as meaning the descendants of a single group of populations, descendants of a single species' (Mayr 1942, p. 280), because in that sense any two groups could be combined in a 'monophyletic' group by going back far enough to their common ancestor (Hennig 1966, p. 72). It is, however, easy to add to Mayr's definition the requirement that the ancestral species itself must be a member of the group. Then a monophyletic group is simply one whose common ancestor would be classified as a member of that group, to be contrasted with a polyphyletic group, whose common

The techniques and justification of evolutionary taxonomy

ancestor would not be classified as a member of the same group. Such is the first important meaning of the word. Reptiles and birds are probably monophyletic in this sense. Some evolutionary taxonomists then worried that their aim to classify monophyletic groups was violated by groups such as mammals, which (it was thought) were polyphyletic by the definition (Simpson 1959a). Simpson (1961) duly redefined monophyly to save the phenomenon. 'Monophyly is the derivation of a taxon through one or more lineages from one immediately ancestral taxon of the same or lower rank' (Simpson 1961, p. 124). By changing 'ancestor' to 'ancestral taxon', mammals could be made into a 'monophyletic' class even though they are (it was then thought) derived from several reptilian groups, because the relevant reptiles came from a taxon of lower than class rank. But the definition only thinly disguises its philosophical error, which is to allow artificial, convergent groups past the definition of monophyly, and into evolutionary classifications. Mayr (1969) accepted Simpson's definition, and duly altered the glossary of his book from the vague formula of Mayr, Linsley, and Usinger (1953). But I wish only to mention Simpson's definition. The other important definition of monophyly, besides that labelled above the 'first important meaning', states that a monophyletic group contains all the descendants of a common ancestor. It is Hennig's (1966) definition. The difference between the two important definitions is that the evolutionary taxonomist's admits groups like reptiles, which are excluded by Hennig's (Figure 2.1). Hennig first published his definition in 1950, and it has since been approved by Sokal and Sneath (1963), most of Hennig's followers, and, I think, an increasing number of other taxonomists. But it is not approved of by Mayr (1969, p. 75), who believes it is 'completely contradicted by common sense' and 'in opposition to the phenomenon of evolutionary divergence'. Evolutionary taxonomy is certainly tied to the former, broader

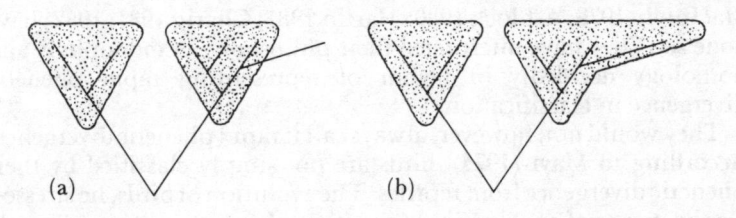

Figure 2.1
Two different meanings of monophyly. (a) The evolutionary taxonomist's meaning: a group is monophyletic if it shares a common ancestor that would be classified in that group. (b) The cladistic meaning: a group is monophyletic if it contains all the descendants of a common ancestor. Not all the descendants of the common ancestor need be included in a group for an evolutionary taxonomist to call it monophyletic.

Evolution and Classification

meaning. It does not define groups that are monophyletic in Hennig's sense. With the broader meaning a group can be called monophyletic even if it lacks some of the descendants of its common ancestor: the Reptilia can be defined and termed monophyletic. It is defined, moreover, by ancestral homologies. Ancestral characters reveal whether a group shared a common ancestor: it is hardly surprising that evolutionary taxonomists put ancestral homologies to use.

Evolutionary taxonomy has practically interpreted its philosophy of representing evolution as the aim to classify monophyletic groups, recognized by homologies. Homologies define monophyletic groups in the sense that they share a common ancestor that possessed the homologies; but if some of the descendants of that ancestor then lost the homologies, they will not be included in the group. The interpretation leads to a compromise among aspects of evolution, but a principled compromise, of a particular kind: phenetic resemblance can override the pattern of splitting; but only for phenetic resemblance in ancestral homologies. The kind of phenetic information that the evolutionary taxonomist will not admit (if we ignore Simpson's definition of monophyly) is convergent similarity. Convergent characters do not reveal common ancestry: they therefore must be excluded.

But in the original treatises that I cited, the conflict between rates of divergence and order of splitting was not taken to be a crucial issue. It was noticed as a difficulty, but not as one of great importance. They were content with some uncertainty as to whether splitting or phenetic divergence should be preferred in particular cases. Since then, however, the cladistic school has appeared. This has forced evolutionary taxonomists to state their preferences more boldly, and both Mayr (1974, 1981) and Simpson (1975, also 1978, pp. 271, 275), as well as others (e.g. Johnson 1970, p. 233; Bock 1974; Gould 1977b; Hecht and Edwards 1977; Szalay 1977; Halstead, White, and MacIntyre 1979; Ashlock 1980; Martin 1981; Charig 1981, 1982) have done so. They have interpreted their philosophy of monophyly and homology decidedly in favour of representing rapid phenetic divergence in classification.

They would not, however, always call it rapid phenetic divergence. According to Mayr (1974), birds are not simply classified by their phenetic divergence from reptiles. The evolution of birds, he stresses, was by means of an adaptive innovation: the change was not simply phenetic, it was adaptive; and the evolution of a new adapted type should be classified as a new 'grade' (Huxley 1958). Thus: 'Rensch, [etc.] ... have particularly emphasized the importance of these levels of adaptation, designated by Huxley as *grades*. All members of a grade are characterized by a well-integrated adaptive complex' (Mayr 1974 [1976, p. 450]). And 'In the history of the vertebrates we know many such cases of the formulation of successful new grades, such as the sharks, the bony fishes, the amphibians, reptiles, birds, and mammals. Each of these is characterized by a certain type of adaptation to the environment' (ibid.). Other revolutionary taxonomists agree (Simpson 1953, p. 346; 1959b, p. 267; 1961; Bock 1965).

The distinction of adaptive from phenotypic change matters because, as the next chapter will show, phenotypic similarity is an ambiguous concept. The status of the concept of adaptive similarity is less certain. It may be ambiguous as well, or it may be capable of unambiguous definition; I would only remark that the matter has yet to be usefully considered. Simpson and Mayr may say that such groups as birds and mammals are characterized by a well-integrated adaptive complex; but before the concept can be discussed it must first be defined. It has not been, at least in satisfactory form. Discussion cannot really even begin, let alone conclusions be drawn. Classification by shared adaptations is a separate principle from the two – phenetic and phylogenetic – we have been concerned with. For reasons that will become clear (pp. 99–106), I call it teleological classification. If we were to group species by their grades, we should need techniques to recognize shared adaptations, which yield unambiguous and hierarchical relations between species according to the similarity of their adaptations. We scarcely know whether species possess unambiguous adaptations, and we certainly lack any formal techniques to classify them. Teleological classification is a problem for the future. But that is not an end to the difficulties. Evolutionary taxonomy does not propose to group species purely teleologically by their grades. It would classify by evolution in some cases, and grades in others. The two concepts are quite distinct, and may even contradict each other: according to Simpson (1959a,b), the mammalian grade is polyphyletic; and even if this case would now be denied (Crompton and Jenkins 1979; Kemp 1982a), a grade could still theoretically be polyphyletic. Evolutionary taxonomists have yet to explain why their philosophy demands a phenetic (or teleological) classification in the case of rapid phenetic divergence (or adaptive breakthrough), but not in the case of convergence. Even if the technical problem of recognizing shared adaptations is solved, the problem of when to classify into grades and when phylogenetically will remain. Major groups may arise by adaptive innovations, but so too do species; the difference is a matter of degree (Simpson 1953, p. 347). Evolutionary taxonomy must specify where on the continuum of degrees of adaptive change it would abandon phylogenetic classification, and classify by grades. Moreover, the specification must be principled, or the whole system will collapse into subjectivity. It would, I think, be very difficult to find any such principle; unless, perhaps, macroevolution and microevolution are different in kind (Goldschmidt 1940; Gould 1980). Both the

Evolution and Classification

teleological principle and evolutionary taxonomy's use of it are too informal. Shared adaptations are practically recognized mainly by phenotypic similarity, and if evolutionary taxonomy recognizes separate groups characterized by 'adaptive breakthroughs', it does little more than classify according to phenetic divergence.

The original justification (as we saw) for doing so was simply that phenetic divergence is a real part of evolution, which should be represented in an evolutionary classification. For now, I am only pointing out that the justification is not consistent, for convergence is a real part of evolution too; but stronger criticism will come later. We have also to deal with two more justifications, which have been advanced more recently: humanity and information content. It is inhuman, some say, to dismantle such groups as reptiles and fish, which are 'more in tune with straightforward human perception' than groups defined only by the order of splitting (Martin 1981, p. 130), and so obviously real (Halstead 1978; Halstead, White, and MacIntyre 1979). I have no more to say about this, because it is too far removed from my theoretical concerns. The other argument is that a classification incorporating two kinds of criteria, phenetic and cladistic, is more informative than one with only one of them. We shall come back to that.

In summary, the evolutionary taxonomist tries to classify species according to their evolution. Convergent similarities of appearance are vigorously excluded: their exclusion is the 'first step' in the system. The main method is the comparison of characters, to distinguish taxonomically good characters, which are those likely to be homologous, from taxonomically bad ones, which are those more likely to be analogous. Good characters are weighted highly in the classification, bad ones lowly: the classification represents monophyletic groups, defined by homologies. The system seems to have two main difficulties. Its philosophical aim to represent evolution is ambiguous: although it has been interpreted unambiguously, it remains to be seen whether the interpretation is valid. And secondly, its methods of distinguishing homologous from analogous characters, although theoretically coherent, are practically imprecise and permit subjectivity. The two more recent schools of taxonomy share a dislike of the subjectivity of the methods of the evolutionary school. But whereas the phenetic taxonomist's reaction is to remove them, the cladist's is to reform them. Phenetic taxonomists remove the whole difficulty by ceasing to try to represent evolution in classification; once that is accepted it is possible to develop more reliable techniques with conceptually more modest aims. The phenetic taxonomist would remove from classification all cladistic information. The cladist's decision is the exact opposite. Cladists would remove all the non-cladistic, phenetic information; their task is to improve the techniques of detecting evolution.

3
The techniques and justification of phenetic taxonomy

Phenetic taxonomy, which I shall here identify with the numerical taxonomy of Sokal and Sneath (1963), can be understood as a reaction to the character weighting of evolutionary taxonomy. We have seen how character weighting is fundamental to evolutionary taxonomy, but is only imperfectly formalized. When the phenetic taxonomist observes the vague weighting criteria of evolutionary taxonomy, he does not like what he sees. He has a precise mind, a taste for measurement and computers, and finds little pleasure in what evolutionary taxonomists so unblushingly call an art. But he has a solution. All the artistry is to be swept aside. If, he says, there are no objective techniques of weighting, then there shall be no weighting; and if the only point of weighting was to represent evolution in classification, then, he says, something else will be found for classifications to represent. In numerical phenetic classification, all the characters are measured, equally weighted, and aggregated in a quantitative measure of the similarity of the taxa under investigation. The classification then represents not evolution, which the phenetic taxonomist thinks in any case is undetectable, but what it has always really represented, if in an imperfect form: phenotypic similarity.

The justification of phenetic taxonomy follows directly from its reaction to the character weighting of evolutionary taxonomy. Classification could in theory be either phylogenetic or phenetic, but because Sneath and Sokal (1973, pp. 53–60) and other pheneticists (Cain 1982) judge the methods by which phylogenetic relations are determined to be too uncertain for practical use, or even circular (see above, p. 27), practical classification, they say, ought to be phenetic. Phylogenetic divergence took place in the past; and phylogenetic relations cannot be known with the certainty of simple observation. Therefore, phenetic classification, which its advocates sometimes call 'empirical' taxonomy, would rely on quantitative, operational, untheoretical measurements of characters: leg length, arm length,

Evolution and Classification

chaetal number. There would be no need for subjective assessments. All the characters, once measured, would be weighed equally in the classification: all characters would be equally good. The whole procedure (they believed) would be perfectly objective and repeatable; from the same starting point, any taxonomist would inevitably arrive, in the end, at the same classification. This happy result could be contrasted with the endless disagreements of evolutionary taxonomy, which were only to be expected given the subjectivity of its techniques. Thus, if we turn to the section of Sneath and Sokal (1973, p. 11) entitled 'the advantages of numerical taxonomy', we find two main, related words – repeatability and objectivity.

So, repeatability and objectivity were the aim of numerical taxonomists. And they devised statistical techniques to realize it. I may remark that the development of statistical techniques to compare phenetic resemblance is uncontroversially an important contribution of the phenetic school. I am not, however, going to spend long on the techniques or go into detail (which is abundantly supplied in their book). We are concerned with the controversial part of phenetic school, the assertion that statistical comparisons of phenetic resemblance not only should be made, but (because of their objectivity) should be used to classify. We shall only need to examine the techniques in sufficient detail to discover whether they truly are objective: further review would be superfluous here.

How, then, is a numerical phenetic classification produced? It is what Sneath and Sokal call 'the fundamental position of numerical taxonomy' (1973, p. 5) that 'classifications are based on phenetic similarity' and 'a priori, every character is of equal weight in creating natural taxonomy'. The procedure of classification begins with plenty of phenetic measurements, of the more characters the better (leg length, nose length, hair length, immunoglobulin concentration, and so on), in each of the taxa under investigation. These taxa they term, in the rude idiom of a barbarous philosophy, 'operational taxonomic units', but that is too much to swallow: we shall term them, for sake of example, species; they could, of course, in practice be any Linnaean level, or just organisms.

The techniques operate on numbers called distances, which can be calculated, for any pair of species with respect to any one character, by the subtraction of the measured value of the character in one species from that in the other. If the measurements for two characters are plotted on a two-dimensional graph, there will again be a distance (the Euclidean distance) between the two points for any species. Likewise there is a distance between two species in the 102-dimensioned plot of 102 characters; but 102 dimensions are difficult to draw. We shall confine ourselves to the two-dimensional case: exactly the same points apply in higher dimensions. Thus, if we have measured two variables, say immunoglobulin concentration and

The techniques and justification of phenetic taxonomy

body length, in each of five species, an average point can be plotted for each species (Figure 3.1). Phenetic taxonomy then converts the measurements into a hierarchical classification, by means of a cluster statistic. We shall use, for example, a 'single-linkage' (or 'nearest-neighbour') cluster statistic. The method is to move out from each point, and as more and more distant species are encountered, steps are formed in the hierarchy. The nearest pairs are 1 and 2, and 3 and 4,

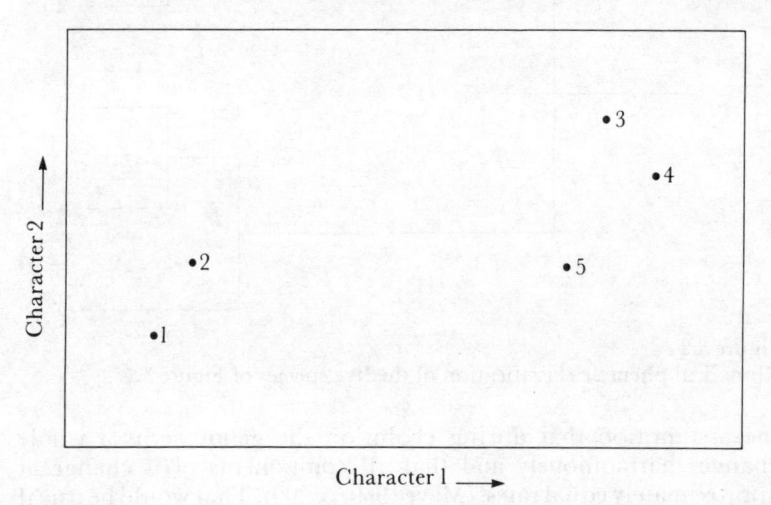

Figure 3.1
Five species, measured for two characters: body length and immunoglobulin concentration. The same argument would apply (in the text) if the axes did not represent single characters, but statistical aggregates of the measurements of many characters.

each separated by a distance of one unit. These are the first steps. At 1.8 units the cluster of 3 and 4 captures 5 (whose nearest neighbour is species 4, 1.8 units away). The two clusters are not joined until we arrive at 5 units, to give the final classification (Figure 3.2).

We have now seen the method, and (in a simple example) the resultant classification. Two points should be made before we move on. The first is that the classification is purely phenetic. It does not attempt to represent phylogeny. Evolutionary convergence is quantitatively represented in the classification as taxonomic similarity. It is no use objecting that convergence should not be allowed to obscure the relationships of species in the classification, because numerical taxonomists completely deny that it is an obscurity. They are not trying to represent, or discover, phylogeny: they are trying to represent something which is to them more easily known than phylogeny: simple phenetic similarity. It is not true that 'equal weighting and random selection of characters are implicitly based on

Evolution and Classification

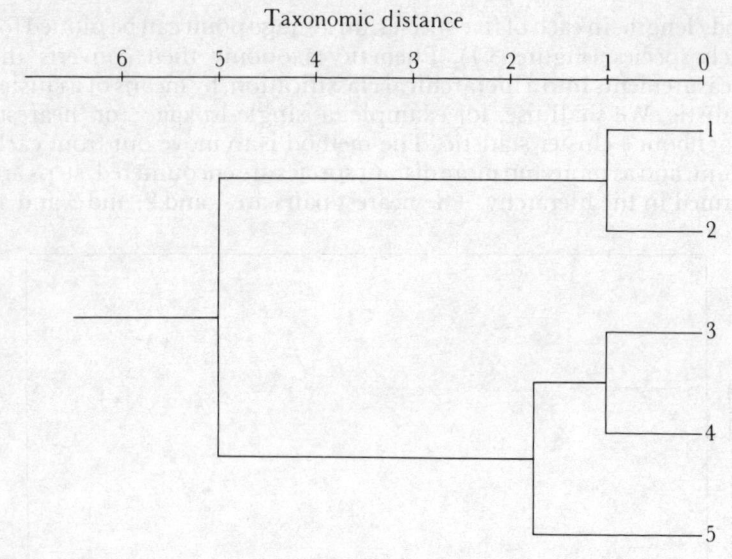

Figure 3.2
Numerical phenetic classification of the five species of Figure 3.1.

the assumption that during evolution the genotype as a whole changes harmoniously and that all components of it change at approximately equal rates' (Mayr 1969, p. 209). That would be true if the technique was intended for evolutionary taxonomy. However, it is not; it does not make any assumptions about evolution. It is intended only to discover phenetic similarity. Likewise, although sibling species should properly be distinguished in evolutionary taxonomy, if they really are phenetically indistinguishable they should not be in a phenetic classification. It is therefore not a criticism to say that 'sibling species are not perceived by numerical taxonomy, whereas some race differences may loom unduly large' (Dobzhansky 1970, p. 358; also Mayr 1969, p. 208). That is what the method aims to do. It is not a criticism of a school to point out that it successfully fulfils its declared intentions.

The second point is that the procedure is, as the numerical taxonomist claims, perfectly repeatable. The whole procedure is exactly specified, and any two taxonomists who started from the same point would finish with the same classification. If they started with the same measurements they would arrive at exactly the same hierarchy. If each were left to choose which characters they were going to measure, the exact hierarchy might differ, but not by much if the number of characters was large. Although a classification made with measurements of only immunoglobulin concentration and body length would probably differ greatly from one made with measurements of body temperature and the number of neurones in the brain, if instead hundreds of characters had been used in each case, then even if there were little overlap between the two sets of hundreds of characters, the final hierarchies would probably be very similar. In that sense the techniques of numerical taxonomy are repeatable.

But we must now look less sympathetically at the 'objectivity' of numerical taxonomy: we must turn to the arguments of a paper of acknowledged importance by L. A. S. Johnson (1970, first published in 1968; Sokal 1973, p. 339, for instance, described it as *'the* outstanding critique of numerical taxonomy ever written'; Wiley 1981, pp. 79–82 gives a different treatment of the same subject). Johnson discussed several kinds of inherent subjectivity in the procedures of numerical taxonomy. Subjectivity enters right at the start, with the choice of which characters, out of the infinity of possible characters of an organism (Johnson 1970, p. 213), are to be measured; and it can then be found at all succeeding stages, the choice of how the character is to be described, how 'distance' is to be measured, which cluster statistic is to be used. We shall concentrate on one, the last-mentioned, of these; it is especially important because it is a ground on which the other schools of taxonomy, as we shall see, can claim an advantage.

When we considered the technique of numerical taxonomy I simply picked a 'single-linkage' cluster statistic, and applied it as if there were no difficulty in the choice. In fact there is a difficulty. It is a difficulty which is little discussed by numerical phenetic taxonomists, who treat the choice of cluster statistic as, at worst, a practical problem. But it is more than that. It is a difficulty of fundamental philosophy, and is profound enough to destroy their claimed advantage. The difficulty is that there is no natural aggregate measure of similarity among species.[1] Of course, techniques do exist, and the single-linkage, or nearest-neighbour, technique is one of them; the difficulty is that there is more than one, but no criterion (within the phenetic philosophy) by which one may be preferred to another. The point can be demonstrated by working through a particular example. Suppose that we have measured, in seven species, two different characters, and that the measurements look like those in Figure 3.3. (Again, the point is the same in 102 as in 2 dimensions, but the latter is easier to draw. In fact, this particular difficulty of phenetic classification deepens as more dimensions are added.) If we again follow through the single-linkage statistic with which we are familiar, species 1, 2, and 3 soon form one cluster, and species 5, 6, and 7 another. Species 4 will first be captured by the cluster of 5, 6, and 7

---

[1]Like Charig (1982, p. 371), but for a different reason, I prefer the term 'aggregate similarity' to 'overall similarity'. Like him, I shall use the term we prefer. 'Overall similarity' is the usual term.

Evolution and Classification

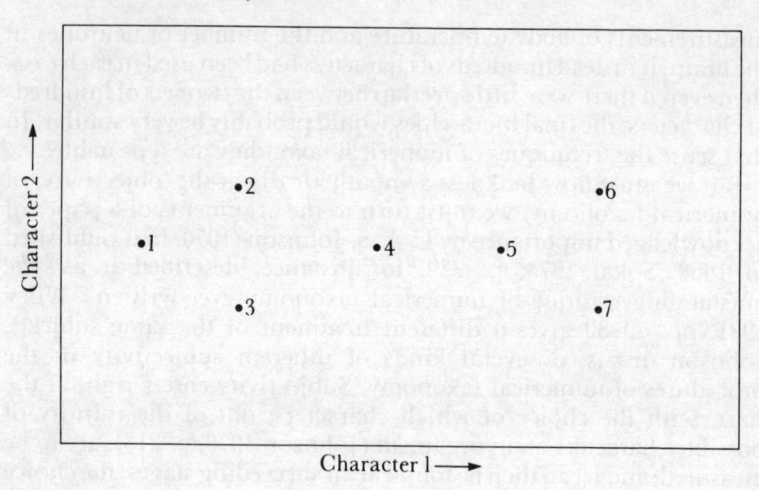

Figure 3.3
Seven species, plotted in a two-dimensional taxonomic space. The axes could represent measurements for single characters, or the statistical aggregates of many characters.

because species 5 is its nearest neighbour. The resultant classification is Figure 3.5b.

The 'single-linkage' statistic, however, is just one of many cluster techniques. We could instead have used a 'furthest-neighbour technique in which a species is joined to the cluster to which it has the nearest furthest neighbour than the nearest nearest neighbour. Or, instead of taking that cluster with the nearest or the furthest neighbour, we could calculate the average distance from a species to a cluster, and join the species to the cluster which was the shortest average distance away. In Figure 3.4 I have marked the average and nearest linkages from species 5 to the two clusters that are competing for it. Cluster B has the nearer nearest linkage, but cluster A has the nearer average linkage. If we use the average-linkage statistic we obtain one classification (Figure 3.5a); if we use the nearest single linkage we obtain a different classification (Figure 3.5b).

The hierarchy, in short, depends on the cluster statistic. This point is general. It does not apply only to single-linkage and average-linkage techniques. There are many more cluster statistics, and they all produce their own kinds of hierarchies. Sneath and Sokal (1973, pp. 204–53) have discussed dozens of different cluster statistics – principal-components analysis, factor-analysis, discriminant-function analysis, single-, complete-, and average-linkage statistics, and so on – and an infinity can be imagined. The reader can turn the pages of *Numerical Taxonomy* and observe, in the illustrations conveniently provided, the different hierarchies produced by each different statistic

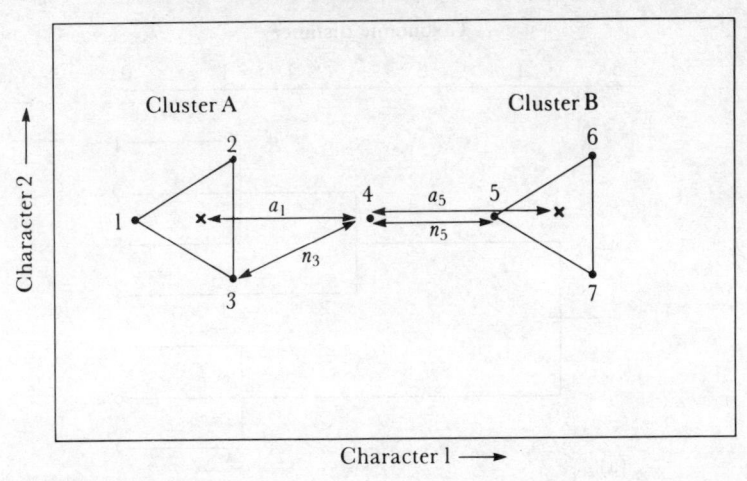

The techniques and justification of phenetic taxonomy

Figure 3.4
Conflict of cluster statistics. Two clusters, of species 1, 2, and 3, and of species 5, 6, and 7, have formed: which cluster should species 4 join? Its nearest neighbour is species 5 ($n_5 < n_3$); but its average distance to the cluster (A) of 1, 2, and 3 (a_1) is less than that to the cluster (B) of 5, 6, and 7 (a_5). Thus a nearest-neighbour cluster statistic species 4 to one cluster; an average-neighbour statistic to the other: the resulting classifications differ (see Figure

from the same raw set of measurements. And the point is not only theoretical. Johnson (1970, p. 223) provides an illustrative list of references to numerical taxonomic papers in which various authors have obtained different classifications of real organisms from the same data.

The choice of cluster statistic is not the only problem. Johnson discussed several others, of which we may mention one more: the measurement of distance. We have so far been clustering according to the Euclidean distance between species. In two dimensions, for two species, the Euclidean distance is the hypotenuse (calculated by Pythagoras's theorem) of the triangle of which the other two sides are the distances apart of the two species in each dimension. But this is not the only measure of distance. One of the early works of numerical taxonomy, by Cain and Harrison (1958), suggested that the distance between species should be measured by 'mean character distance' (MCD). The mean character distance is the average of the distances of each dimension (rather than the sum of their squares). As with different cluster statistics, so with different measures of distance, the hierarchical classifications may differ according to which is used.

Although the different techniques have been classified and their different properties discussed – some produce short hierarchies with

Evolution and Classification

Taxonomic distance

(a)

(b)

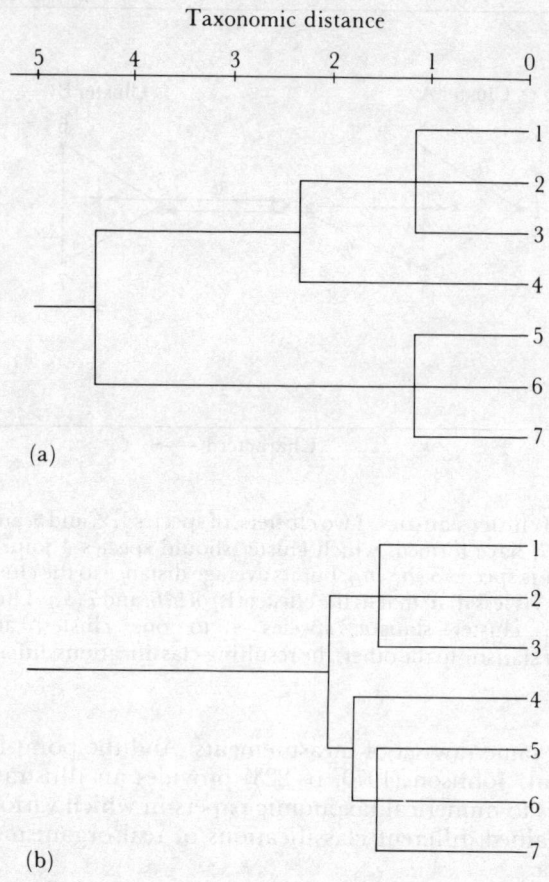

Figure 3.5
Numerical classification of the seven species of Figure 3.3: (a) by average-neighbour statistic; (b) by nearest-neighbour statistic. They differ topologically: species 4 is classified differently by the two statistics.

all the species similarly related, others longer hierarchies with more steps – no one has discussed the real relative merits of the techniques from fundamental principles. We do have discussions of computational time, of algorithmic consistency; quantitative methods have been developed to compare different statistics (Sneath and Sokal 1973, pp. 275-90; Rohlf 1974; Rohlf and Sokal 1981). But the question of principle has not been faced. There are two reasons. One is that the phenetic taxonomist feels no strong need to: if phylogenetic classification is impossible, classification can only be phenetic; and Sneath and Sokal (1973, p. 424) are accordingly untroubled by Johnson's demonstration that there is no optimal phenetic classification. The other is that they have not, because they cannot. There are absolutely no criteria to support such a discussion. But although there is no fundamental reason within phenetic taxonomy for preferring one statistic to another, the practical phenetic taxonomist has to use one of them. He will have to make a choice. It is of course perfectly possible to do so, but the choice, when it is made, will be ... no, surely not?

It will be completely subjective. The reason is easy to see. The only objective method of deciding between cluster statistics would be by reference to some higher criterion, which the statistics are actually meant to be discovering. But no such criterion exists. If there were, in nature, a true hierarchy of phenetic similarity among species, then it would be easy to test the different statistics against the hierarchy, find out which one was best at discovering it, and then preferentially use that statistic. In fact, the only measures of phenetic similarity are provided by our statistics; there is no independent, external criterion against which we can test them. For this reason, Johnson (1970, p. 214) stated that there is an inherent circularity in the philosophy of phenetic taxonomy. The cluster statistic is meant to measure aggregate morphological similarity, but that does not exist independently of the cluster statistic: there are as many concepts of aggregate morphological similarity as there are cluster statistics. When the choice is made, even if it is concealed by the circumlocutions of information technology, it will have to be made on practical considerations, or may even be quite arbitrary. In the late 1950s the numerical taxonomists came in exploding petards beneath the subjective old taxonomists they were going to eliminate, only to be hoist, in their turn, and within a decade, on their own favourite device.

The subjectivity of phenetic taxonomy has two broader consequences. The first concerns the evolutionary school, which is damaged by the same criticism as phenetic classification. Evolutionary taxonomy does contain a phenetic element, and (as we have seen) insists that it is real and merits classification. Evolutionary classification is not as purely phenetic as numerical taxonomy, because it excludes convergence; but it does include differential rates of divergence. How can differential divergence be measured? Any attempt to measure it as simple phenetic change is bound to run into Johnson's difficulty, as Johnson (1970, p. 233) himself realized. If it is to be thought of not as phenetic, but as adaptive change, we need techniques to measure adaptive similarity, and techniques that avoid the ambiguity of purely phenetic techniques. No such technique has ever been suggested, although one may be possible. The teleological principle might then supply the need for some theory by which to choose among cluster statistics. Ecological adaptation might be the property of the real world we could check our statistics against. Evolutionary taxonomy, however, has more pressing problems of subjectivity, and discussion does not customarily reach this far. But the evolutionary taxonomist's troubles with weighting are in principle soluble; one day they may be solved. If they ever are, another philosophical difficulty will be expecting the attentions of that school.

The second consequence concerns the distinction of natural from artificial classifications. Biologists disagree over the formulation of this distinction; but, according to one popular form, in a natural classification the members of a group agree in other characters besides the one (or more) used to define it, whereas in an artificial classification they do not. Because the distinction mentions only resemblance of characters, it is phenetic. So, just as phenetic similarity cannot be objectively measured, nor can a single 'most' natural classification be identified. Is the classification of Figure 3.5a more natural than that of Figure 3.5b? The question is unanswerable. The degree of naturalness of a classification cannot be used to judge its objectivity.

This conclusion is strictly circumscribed. It does not mean it is nonsense to discuss the naturalness of a classification, nor that our ideas of phenetic similarity are an illusion. They are both concepts with a meaning. The members of a group, such as the Reptilia, which is phenetically defined, do appear more similar to each other than to other species. In most respects a lizard and a crocodile look more like each other than does either to a bird. It has not been the purpose of this chapter to argue they do not. Phenetic measurements can still be made, as indeed they must be for any evolutionary study of particular characters; the naturalness of classifications can still, at least informally, be compared. We shall continue to use both concepts when their subjectivity and informality do not matter. What the conclusion does mean is that there is no *single* meaning of phenetic similarity or naturalness that could underwrite an objective system of classification. Phenetic similarity alone cannot supply a consistent, unambiguous, objective, hierarchical classification of all life. Lizards and crocodiles may resemble each other more closely than do either with birds: but the implicit measurements and weightings would not necessarily give a hierarchical classification of all life; nor would they necessarily be the same in the case of the gross phenetic similarity of some other animals, such as millipedes, centipedes, and insects; nor would the implicit cluster statistics in these different cases necessarily be consistent. And even if our intuitive ideas of phenetic similarity did supply a unique, consistent, hierarchical cluster statistic, that statistic would still be (in the sense of this book) subjective: it would owe its power to mass agreement, not natural principle. We certainly can measure the phenetic similarity of species with respect to characters; we can aggregate the measurements in multivariate comparisons, if we have some reason to do so; but that does not make

The techniques and justification of phenetic taxonomy

phenetic classification objective. If it is objective, one particular measurable cluster statistic must represent a unique natural hierarchy. The argument we have been through suggests that, within the phenetic system, that is impossible.

I have one final point, which will take us on to the next chapter. Phenetic taxonomists, as we have seen, oppose the weighting of characters. They dislike the subjectivity of the methods of character weighting in evolutionary taxonomy. But in fact, if an explicit, principled, quantitative case can be made for weighting, there is no real reason (within its philosophy) why numerical taxonomy should not admit cluster statistics that systematically weight some kinds of characters heavily, and others lightly. When I quoted earlier from Sneath and Sokal, they said numerical taxonomy was only opposed to weighting a priori: it could be introduced if a case is made; and they discuss statistics that do so. If only evolutionary taxonomy could provide a quantitative rule to govern its technique, it could then be incorporated into phenetic taxonomy. Phenetic taxonomy could become, in the technical sphere, a kind of evolutionary taxonomy, or vice versa. If only there could be found within evolutionary taxonomy, what cannot be found in the original form of numerical phenetic taxonomy – a real hierarchy for the cluster statistic to aim at – then evolutionary taxonomy could solve the problem of the cluster statistic. In fact, there is in evolutionary taxonomy a natural hierarchy that can serve this purpose – it is the hierarchy of phylogenetic branching. For this reason, numerical taxonomy may be led, by its own logic (Farris 1977, 1979; Mickevich 1978), to develop itself into the school of classification that provides the subject of our next chapter. It is led to become cladism.

4
The techniques and justification of cladism

Justification

The philosophy of phenetic taxonomy proved inadequate to justify any single taxonomic technique. It aimed at repeatability; but the criterion of repeatability alone defines an infinity of classifications, the choice among which is subjective. Hennigian cladism is designed to avoid that fate. Its philosophy allows only one classification: the classification defined by the hierarchy of phylogenetic branching.[1] If two species share a recent common ancestor they are classified together, regardless of their phenetic similarity to other forms. Hennig justified the system by an argument similar in style to this book: he asked which classificatory system was objective. In the first chapter of *Phylogenetic Systematics*, he, like many others, noticed that there are two main possible principles of classification, phenetic and phylogenetic. The question, of course, is which one is better. Hennig tackled it by asking which one sought to represent a natural relationship that is truly hierarchical. He reproduced a figure, which he took from Gregg, of a true hierarchy (see Figure 4.1); and asked which principle identifies a natural relation of the figured kind. He took phenetic similarity first, and showed that it does not give hierarchical classifications of the form of Figure 4.1. He actually directed his argument against the idealistic morphology that prevailed in interwar Germany, and it is not clear how well he understood the difficulty the later more sophisticated numerical phenetic taxonomy would face. 'The form relationships between groups of morphological systems in the broadest sense are not exactly

---

[1] I shall follow Hennig, and use 'phylogeny' here to mean only the order of branching of a set of species, not their rates of evolution between the branching points; phylogenetic relations here are only those of ancestry, not of phenetic similarity. Mayr (1969, 1974) disagrees with this usage.

The techniques and justification of cladism

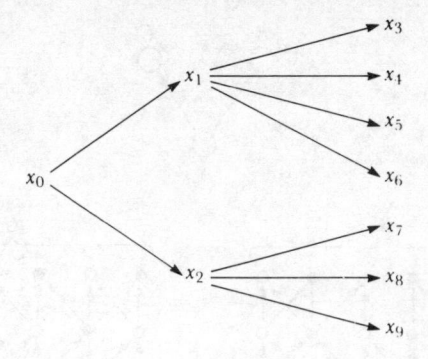

Figure 4.1
The relations between entities in a true hierarchy. Any identifiable property of nature that has such a structure could provide the principle of an objective system of classification. (From Hennig, W. (1966), *Phylogenetic Systematics*, with permission from The University of Illinois Press.)

measurable, since there is no known method of measuring similarities and differences of form. Many believe [that] mathematical bases for exact measurements of similarities and differences in form will eventually be found, but we are skeptical' (Hennig 1966, p. 23). Sneath and Sokal (1973, p. 54) confidently replied that 'the development and success of numerical phenetics has invalidated that argument', but it all depends on what Hennig meant. If he meant that the multidimensional measurement of form is impossible, they would be right; but not if he had something in mind more like Johnson's argument which we considered in the last chapter. Hennig's remarks elsewhere (1966, pp. 74-5) suggest the latter, but his point stands either way. The phenetic philosophy is not objective; phenetic relationships are not unambiguously hierarchical.

So phenetic classification is ruled out. What about the other possibility, phylogeny? The branching pattern of phylogeny is an unambiguous, natural hierarchy. In Figure 4.2 I have reproduced Hennig's figure 4, which illustrates the actual pattern of evolution and thus demonstrates that the branching hierarchy of phylogeny has exactly the form of a true hierarchy (Figure 4.1). 'The structure of the phylogenetic relationships that must exist between all species according to the assertions of the theory of descent is necessarily that shown in Figure [4.1] ... Therefore the species of biological systematics can be substituted for x_0, x_1, x_2 ... in Figure [4.1]' (Hennig 1966, p. 20). Phylogenetic relationships really are hierarchical. They are also unambiguous, because a set of species either do share a unique common ancestor or they do not. If we have three species, then it either is or is not true of any two of them that they share a more recent common ancestor with each other than with the third species;

47

Evolution and Classification

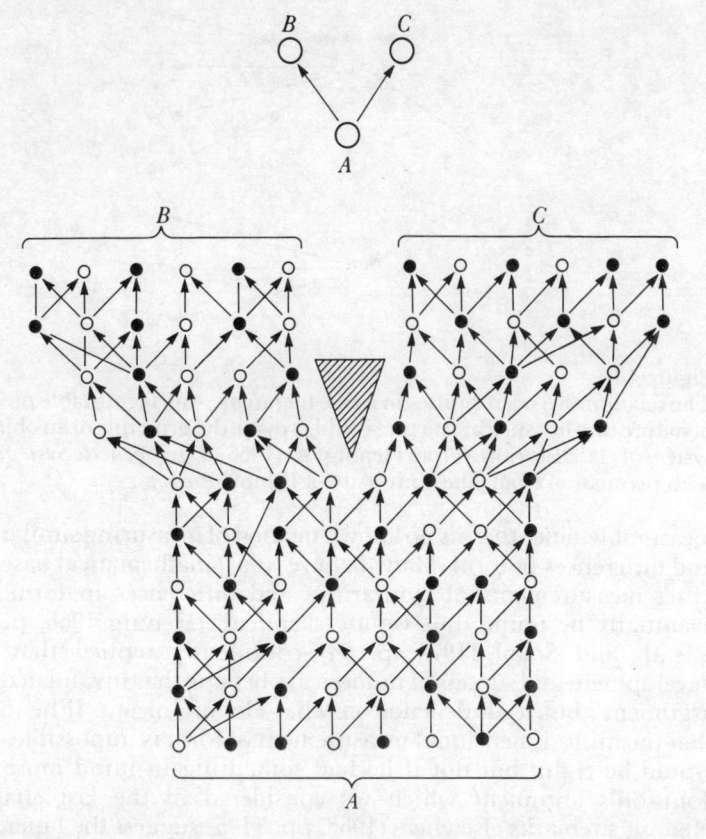

Figure 4.2
Hennig's branching pattern of phylogeny. The ancestral relations of species have the hierarchical structure of Figure 4.1, because species split in this unambiguously hierarchical pattern. (From Hennig, W. (1966), *Phylogenetic Systematics*, with permission from The University of Illinois Press.)

if they do, on the phylogenetic principle they should be grouped together. The definition of groups is unambiguous: a group should only be classified if it has a common ancestor unique to itself. The same is not true of phenetic classification. Phenetically defined groups cannot generally and unambiguously be substituted for x_0, x_1, x_2 ... in Figure 4.1. The phenetic philosophy does not unambiguously dictate whether any two species should be classified together, because whether a set of species forms a classificatory group in the phenetic system depends on which of the many meanings of 'phenetic' is understood, and different meanings imply different classifications.

The techniques and justification of cladism

There is no single hierarchy of phenetic resemblance, there are as many of them as there are cluster statistics. Phylogenetic classification should in principle be unambiguous, because there is only one hierarchy of phylogeny. If it can be discovered, it will underwrite the only known objective system of classification; and should therefore be the aim of classification. It should provide what Hennig called the 'general reference system' of biology.

A digression on prokaryotes

If a set of species does not possess unique phylogenetic relationships, Hennig's philosophy will not apply to it, and cladistic classification will become ambiguous. All the genes of a set of species with typical Mendelian inheritance have descended through the same phylogenetic hierarchy: the phylogeny of cytochrome genes in vertebrates is the same as the phylogeny of haemoglobins. That is why the phylogenetic classification is unambiguous. If, however, the different genes of a 'species' had experienced different phylogenetic histories, and the species themselves had experienced mixed ancestries, Hennig's argument would break down. The species would not longer possess the unambiguous relations required for them to be 'substituted for x_0, x_1, x_2 ... in Figure [4.1]'.

Prokaryotes may provide an example. The 'species' of prokaryotes may not have unambiguous ancestries, because of the frequency of 'horizontal' gene transfer. Genes certainly do move between prokaryote species. The phenomenon is well documented. It was discovered independently by two groups of Japanese biologists, led respectively by Akima and Ochiai, in the late 1950s; and became well known elsewhere after their work was reviewed in English in 1963 (see Falkow 1975; Broda 1979). Both groups performed the experiment of brewing *Escherichia coli* resistant to a certain antibiotic together with another genus of bacterium, *Shigella*. Resistant *Shigella* could later be isolated from the brew. Plasmids called R factors carry the genes conferring resistance between species. Much is now known about the ease of movement of different R factors between different kind of bacteria (Falkow 1975, pp. 78–9), the rates of which vary from 10^{-3} to less than 10^{-8} depending on the species. No one doubts that the plasmid genes of a bacterial species at any one time have experienced multiple ancestry. If bacterial species were defined by their plasmid genes, Hennig's philosophy could not apply to them.

But the plasmid genes are actually unimportant. They do not determine much of the bacterial phenotype. Most of the information is contained on the bacterium's larger chromosome of DNA, which is as separate from any plasmids as is the nuclear DNA of a mammal from that of a mammal-dwelling virus. The chromosomal genes are

Evolution and Classification

the genes that matter. It is against them that Hennig's philosophy should be tested. If horizontal gene transfer is rare among the genes of the bacterial chromosome, if most of the genes of a bacterial species share the same ancestry, Hennig's philosophy can apply to bacteria just as it does to larger eukaryotes. (For even if genes do transfer among eukaryotic species (Lewin 1982c), they probably do so far less frequently than in prokaryotes.) But if it is common, the phylogeny of the customarily recognized bacterial species would be ambiguous. Hennig's philosophy would then not apply to the bacterial 'species' in the same way as it does to eukaryotes. Indeed, the philosophy would challenge whether bacteria form ordinary species at all: in the extreme, it would suggest, that all those bacteria that routinely exchange genes should be placed in one species, analogous to a reproductive species of, for instance, chimpanzees. The cladistic classificatory|hierarchy of the reproductively defined species would have a reduced number of levels, smaller than in the customary bacterial classifications. A richer hierarchy, featuring the phenetically recognized bacterial species rather than the more inclusive cladistic species, could be defined by a phenetic cluster statistic if there were some reason to do so: it might help in practical recognition, perhaps; for although phenetic classifications are subjective, they are not necessarily useless. A phenetic classification of bacteria would not, however, be of any general interest to the evolutionary biologist, who would wish to know the phylogenetic history of each individual gene.

The important question, therefore, is how often the genes of the bacterial chromosome move between species. Is the frequency high, as in plasmids; or low, as in a typical eukaryote? Whatever it is, it is not zero. Under certain conditions, plasmids may integrate with their bearer bacterium's chromosome; the plasmid may then pick up a gene from the chromosome, become independent again, and if it were then to transfer to another species, and integrate in turn with its host's chromosome, a 'normal' bacterial gene (rather than a plasmid gene) might move between species. But although chromosomal genes can be horizontally transferred, it is a very rare event indeed (Falkow 1975, p. 198); different bacterial species do have characteristic genomes (Cullum and Saedler 1981, p. 146). In bacteria genes may be transferred more often than in other groups (although they may not be), and the concept of species may be more difficult, but they do not move frequently enough to break down the distinctive character of each species' genome. Phylogenetic classification is probably possible in much the same way for prokaryotes as for eukaryotes (Stackerbrandt and Woese 1981). Bacteria probably do not in fact pose any great philosophical puzzles for cladism. Plasmid genes do have a mixed ancestry, and if the plasmids were thought of as part of the total genome of the bacterium, the bacterium would proportionally have a mixed ancestry. However, the plasmid genome is probably better thought of as analogous to a virus in a eukaryote. Viruses move among species, from simian monkeys to humans for example, and if we are to say that bacteria have mixed ancestries because of their R factors, we might just as well say that a person harbouring such a virus has a mixed ancestry too.

According to Sneath and Sokal (1973, p. 54) viruses themselves provide a difficulty for cladistic (or evolutionary) classification. They make two points. One is that bacteriophages and other viruses may have mixed ancestry, just as plasmids do. We have already discussed that: the cladistic philosophy can be applied, but (as we have just seen) it might change what we call a species and classify the prokaryotes into relatively few hierarchical levels; a hierarchical phenetic classification of phenetically recognized species would as usual be subjective. Sneath and Sokal's second point concerns the work of Subak-Sharpe *et al.*, which 'has shown that the pattern of small DNA viruses is very close to that of mammalian DNA, suggesting that these viruses may have arisen from mammalian DNA sequences'. They conclude that 'the only way to bring order into such a system is by a phenetic classification'. I disagree with this. If mammalian viruses are descended from mammals, in the cladistic system they must be classified with them. That is the logical way to bring order into the system. It may sound strange, but the cladistic classification is unambiguous and automatically implied by the school's philosophy. Of course, if mammalian viruses are descended from mammals, snake viruses from snakes, and honeybee viruses from honeybees, the group 'virus' would cease to have any formal classificatory validity. It could be retained as a nonclassificatory group, analogous to the group of 'animals with wings', but if it is not a monophyletic group, there is no doubt how cladism would deal with it; it presents no philosophical difficulty: the taxonomic category 'virus' should be exploded. Phenetic classification is therefore not 'the only way to bring order' into viral classification; cladism is a simple alternative.

So much for prokaryotes. Cladism probably works as well with them as with any other group of organisms. From now on, we can assume that the phylogeny of whatever living things we wish to classify is unique and unambiguous, and concentrate on how it may be discovered. Before we discuss that technical problem, however, I should clear up a small verbal confusion. In a cladistic classification, a species is grouped with that other species (called its sister species), or group of species, with which it shares its most recent common ancestor. The objective relation 'shares most recent common ancestor with' defines unambiguous hierarchical groups. Now, cladists call these groups 'monophyletic', which, as we have seen (p. 30) is a word

Evolution and Classification

with more than one meaning. Evolutionary taxonomists call a group monophyletic if it possesses a common ancestor that would have been classified in the group; the group need not contain all the descendants of the common ancestor. 'Reptilia' therefore are called monophyletic by evolutionary taxonomists. The cladistic meaning of monophyletic is different. 'A monophyletic group is a group of species descended from a single ("stem") species, and which includes all species descended from this stem species' (Hennig 1966, p. 73). For a group to be monophyletic in the cladistic sense it must not only possess a common ancestor, but also contain all the descendants of that common ancestor. Cladists do not recognize the Reptilia, and they have a name for groups of that inadmissable kind: they call them paraphyletic.

Groups, in cladistic language, may be polyphyletic, paraphyletic, or monophyletic (Figure 4.3). Only monophyletic ones are allowed in cladistic classification. The preference (and the meaning of monophyletic) follow directly from the justification of the system. Monophyletic groups are unambiguous branches of the phylogenetic hierarchy; paraphyletic and polyphyletic groups can only be defined phenetically and are therefore ambiguous (Hennig 1966, pp. 73-7). In terms of Figure 4.3, the pattern of cladistically monophyletic groups is clear; but there are many possible and conflicting groupings with paraphyletic groups (species 1 and 2 in one group and 3 and 4 in another, or species 1, 2, and 3 in one group and 4 in another) and the choice among them would have to be made phenetically.

The distinction of cladistically paraphyletic from monophyletic groups is undeniable; but evolutionary taxonomists not unnaturally resented the confiscation of their (as they believe) term 'monophyly'

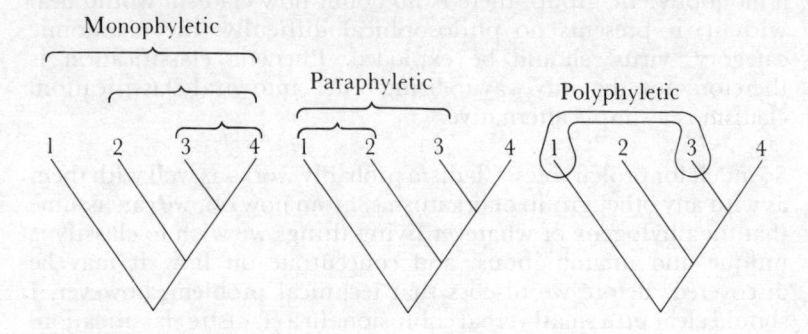

Figure 4.3
The three possible kinds of groups: monophyletic, paraphyletic, and polyphyletic. Paraphyletic and polyphyletic groups are defined phenetically and their group membership is therefore ambiguous. The membership of monophyletic groups is unambiguous.

(Ashlock 1971, 1972; Mayr 1974 [1976 edn], pp. 446–8; Van Valen 1978). Mayr, we have seen, thinks Hennig's definition is 'contradicted by commonsense' and accordingly would retain the term in its broader reference. He would use Ashlock's (1971, p. 65) term 'holophyletic' for groups that cladists call monophyletic. Two main meanings of monophyly do now co-exist in the literature; but the ambiguity need not be a practical problem, because it is usually obvious which is being used. In this work I shall follow Hennig and perhaps the majority of taxonomists: I shall use the cladistic meanings of monophyletic and paraphyletic, and avoid the term holophyletic.

The justification of Hennigian cladism, and its associated terms, is now complete. In summary, there are two possible kinds of hierarchical classification – phenetic similarity and phylogeny. The former aims at an unreal hierarchy and is not objective, whereas the latter is realistic and (by elimination) the only objective system. The next question, after it has been accepted that species ought to be classified by their phylogenetic relations, is whether practical techniques can be devised. If they cannot, cladism will fail. But it will fail, not because it is unphilosophical but because it is unworkable.

What kinds of techniques are needed?

We shall take one section to consider the general form of techniques that are needed, and two more to look at the cladistic techniques themselves. The technical aim is to discover, for each species, which other species it shares its most recent common ancestor with. The evidence available is the phenetic similarity of species with respect to such characters as have been studied. Roughly speaking, more closely related species will resemble each other phenetically more than less related species; but, because some characters are better indicators of phylogenetic relationship than are others, phylogeny can be discovered with greater accuracy by concentrating on particular kinds of phenetic evidence. There are three main reasons why species may share characters. A first division is between homologies and analogies: homologous characters are shared from a common ancestor; analogous ones evolved independently in the different species. The cladist should concentrate on homologies, and ignore analogies, because analogies do not indicate phylogenetic relationships whereas homologies at least can do. The concept of homology decomposes in turn into homologies of ancestral characters and homologies of derived characters. This section will seek to show that only homologies of derived characters indicate shared ancestry. The

53

Evolution and Classification

methods of cladism therefore must be able to distinguish analogies from homologies, and derived character states from ancestral ones; finally the methods must be able to convert the residue of shared derived characters into a classification.

In Chapter 2 I dealt at length with the distinction of homologies from analogies, and I shall not discuss it further in this section. I shall be concerned with the second distinction, of ancestral from derived character states, and with the conceptually easy matter of formal classification. Like most words in taxonomy, 'ancestral' and 'derived' have been used in more than one sense: I must therefore define them. Ancestral and derived here refer only to the successive stages in the evolutionary transformations of characters: when a character changes during evolution, I call its earlier state ancestral, its later state derived. Whether a particular state is ancestral or derived obviously depends on the other states under consideration. Any given character may change many times in evolution. If the skeletal state changed, by degress, from being finless to finned to limbed, then 'finned' is derived with respect to 'finless' but ancestral with respect to 'limbed'. Ancestral and derived are properties of a character, independent of the kind of species they belong to. A more restrictive meaning, not used in this book, should be distinguished. The words have also been used to mean the characters of an ancestral or derived *species*. In that sense, an ancestral character is a character of an ancestral species; if the same character state was found in a descendant species, it would not be called ancestral. In the sense of this book, although an ancestral species would have had ancestral characters, 'ancestral' characters are not confined to ancestral species. An ancestral character may be retained in descendant (derived) species after their evolution from their ancestors; indeed, any character that did not change during this evolutionary event I should call an ancestral character.

The words matter because, as Hennig argued (and others had understood more implicitly), whereas the sharing of a derived character state between two species indicates a phylogenetic relationship, the sharing of an ancestral state does not. Cladists therefore classify groups according to shared derived characters, rather than any old shared characters. The reasoning may be obvious. Suppose we wish to classify a baboon, a crocodile, and a cow relative to each other, and study the states of their limbs. The baboon and the crocodile have five toes, the cow two; but the fact that a baboon and a crocodile both possess a pentadactyl limb – the ancestral condition for tetrapods – is not evidence that these two species are phylogenetically closer to each other than is either to the cow. Shared ancestral characters do not reveal phylogenetic relationships.

If it is not obvious, it can be justified from first principles. Shared ancestral character states are untrustworthy evidence of phylogenetic

The techniques and justification of cladism

relationship because the character may have changed more than once, on different occasions, anywhere in the ramifying branches of phylogeny; the ancestral states may be retained by any combinations of species, regardless of their phylogenetic relations. Derived states, however, will be shared only by the descendants of the particular ancestral species that the new derived state first evolved in (until another change takes place). Consider, for example, a simple phylogeny of four species and the distribution of four characters, each with two states, among the four (Figure 4.4). The ancestor had the four ancestral character states *abcd*. Each character has changed once. Only the derived characters fall into the kinds of groups that cladists wish to recognize: a' and c' indicate cladistic groups (as do, but trivially, b' and d'). The ancestral character states do not generally fall into cladistic groups of species: species 1, 3, and 4 share character state d' but they do not form a cladistically monophyletic group. Only if an ancestral character remains unchanged within a whole branch, and changes at the beginning of the other branch (as character a in Figure 4.4), does it fall into a cladistic group; but that is a special circumstance. Shared derived characters always fall into cladistic groups. They must, because of the way evolution proceeds.

So shared derived characters reveal cladistic groups. The next question is how derived character states may be recognized, and we shall soon come to it. But first I wish to show how the cladist converts a knowledge of the distribution of derived character states into a classification. Actually it is rather easy. Let us classify three species, called 1, 2, and 3, for which we know the states of seven characters

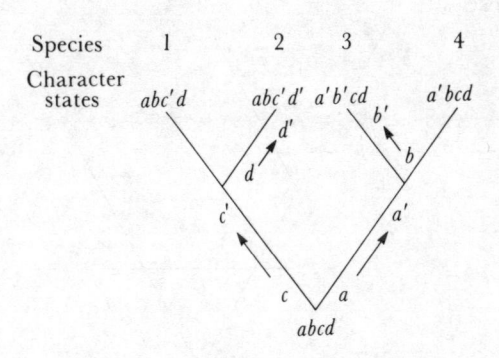

Figure 4.4
Phylogeny of four species, and the distribution of four characters (A, B, C, D) among them. Ancestral states of the characters are simple letters ($a, b,$ etc.); derived states are indicated by primes (a' b', etc.). Only shares derived character states always fall into monophyletic groups; shared ancestral ones may (c in species 3 and 4) or may not (d in species 1, 3, and 4).

Evolution and Classification

(called *A, B, C, D, E, F,* and *G*). We can indicate the ancestral state of *A* by *a*, and its derived state by *a'*, likewise for *B, C,* and the rest. Suppose the character states of the species are as follows: 1 (*a', b', c, d, e', f, g'*), 2 (*a, b', c', d, e, f, g'*), and 3 (*a, b, c, d', e, f, g'*). The cladistic classification (or 'cladogram', Figure 4.5) is made simply by grouping the species according to their shared derived characters. That is all there is to it.

The cladistic classification can differ from the phenetic or the evolutionary classification. The group of three species in Figure 4.5 could be defined by *G*, the group of species 1 and 2 by *B*, species 3 by *D*, species 2 by *C*, and species 1 by *A* (or *E* or *F*). The important points to notice are that the shared ancestral, and the uniquely derived, character states are not used to construct the cladogram. Species 2 and 3 share the same state of *A, E,* and *F*, but they are not for that reason classified any more closely. A phenetic classification would differ; in this case it would probably be the reverse of Figure 4.5. Species 1 and 3 share the same state for one character (*C*), species 1 and 2 share two (*B* and *D*) but species 2 and 3 share three (*A, E,* and *F*). If the species were grouped by their phenetic similarity alone they would probably have been classified as in Figure 4.6. The cladist excludes all similarities except for shared derived character states, and the result is Figure 4.5: the phenetic taxonomist does not distinguish ancestral from derived states, and the result is Figure 4.6: what about evolutionary taxonomy? It is less easy to predict. Evolutionary taxonomists generally choose phenetic classifications when they are suggested by ancestral homologies, but not when they are suggested by analogies.

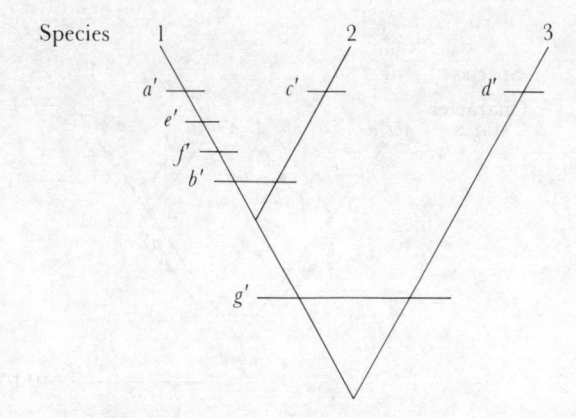

Figure 4.5
Cladogram of three species (called 1, 2, and 3) revealed by the states of seven characters (*a* to *g*). Bars indicate derived characters; a bar across more than one species indicates that the same derived character state is shared by them. Only the derived states are shown: unmarked states are implicitly ancestral.

The techniques and justification of cladism

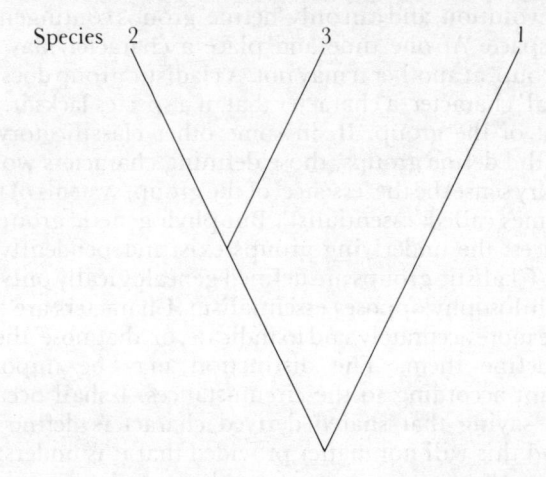

Figure 4.6
Phenetic classification of species 1, 2, and 3 (of Figure 4.5). Note difference from cladistic classification in Figure 4.5.

In this case the similarity of species 2 and 3 is due to ancestral homology, not convergence, and the evolutionary taxonomy would probably be the phenetic one. But even this is not certain: if the uniquely derived character states of species 1 were judged adaptively unimportant in some way, the evolutionary classification might be the same as the cladistic one (Figure 4.5).

A digression on definition

Strictly speaking, the cladistic classification is not *defined* by shared derived characters (Ghiselin 1984). Shared derived characters merely betoken the phylogenetic groups, which have a real, independent existence. The pattern of phylogeny itself defines the classification; shared derived characters are evidence only. At any one time and place a particular derived character may be shared by all and only the members of a monophyletic branch of the phylogenetic tree. It might then seem not only to be evidence of the cladistic group, but to define it as well. However, it does not. If the character immediately evolved into a changed state in a part of a species in the group, perhaps without any speciation event, the cladistic grouping, and cladistic classification, would not be affected. That is because the cladistic classification is technically defined by the underlying phylogeny, not by the evidence used to indicate it. Characters are always liable to

57

Evolution and Classification

change in evolution, and can only 'define' groups contingently upon time and space. At one time and place a character may define a cladistic group, at another it may not. A cladistic group does not have an 'essential' character, a character that, if a species lacks it, rules the species out of the group. If, in some other classificatory system, characters did define groups, those defining characters would (in a classificatory sense) be the 'essence' of the group; systems of that kind are sometimes called 'essentialist'. But phylogenetic groups do not have essences: the underlying groups exist independently of their characters. Cladistic groups are defined genealogically only, and the cladistic philosophy opposes essentialism. Characters are evidence, and may be more accurately said to indicate, or 'diagnose' the groups, than to define them. The distinction may be important or unimportant according to the circumstances. I shall occasionally lapse into saying that shared derived characters define cladistic groups, and this will not matter provided that it is understood as a shorthand, and no one is tempted to read essentialism into it. But the habit is not without danger. Some cladists have come to think that cladistic groups are defined by shared derived characters independently of the phylogeny that the characters are supposed to indicate. Having expunged the distinction between the evidence and what it is supposed to be evidence of, and ruled out the phylogenetic philosophy of cladism, they have fallen, as such people always will (Hull 1965), into essentialism. We shall come back to them.

The question of whether a classification is natural cannot decide among competing classificatory philosophies. But it is still a curious question, and we may ask whether cladistic classifications are natural. The answer in fact depends on the meaning of the question. In a sense, cladistic classifications should be perfectly natural. If we concentrate on shared derived characters – and assume that they have all been perfectly identified – they must fall into exactly the same pattern of groups. Therefore the (shared derived) characters that define the groups should be perfectly correlated with all other, non-defining (shared derived) characters. Characters could only disagree, and the classification be unnatural, if the shared derived characters had been wrongly identified. If, however, we consider all characters indiscriminately, whether ancestral or derived, the cladistic classification will not be perfectly natural. The ancestral characters will disagree with its groupings. Just how unnatural the classification is will depend on the distribution of the ancestral characters. In Figure 4.4 character d disagrees with the cladistic classification. A disagreement with only one character does not make the classification all that unnatural; but it could be that there were one hundred characters like d. The cladistic classification would not be altered, but it would, in the phenetic sense, be exceedingly unnatural.

The techniques and justification of cladism

It is in just such cases that evolutionary taxonomy prefers to recognize groups defined by shared ancestral characters. Reptiles are an example. Reptiles are defined by ancestral homologies, and the cladist unhesitatingly splits off the crocodiles to form a monophyletic group (which Mayr would call holophyletic) together with the birds. Mayr (1974 [1976, p. 436]) thinks he is criticizing cladism, and defending evolutionary classification, as he remarks 'the number of evolutionary statements and predictions that can be made for many holophyletic groups (like birds and crocodiles) is often quite minimal'. He thinks it is a criticism because 'there has long been agreement among the theoreticians of classification that in most cases those classifications are "best" which allow the greatest number of conclusions and predictions' (ibid.) and he even quotes Mill in support of this naive phenetic criterion. As criticism, Mayr's remark has two defects. One is that it is a defence of phenetic, not evolutionary classification. The most phenetically natural classification will be a phenetic one: the main difference between phenetic and evolutionary taxonomy is that the latter excludes convergence; but phenetic taxonomy only admits convergence when it makes a classification more natural. When an evolutionary taxonomist excludes the kind of convergence recognized in a phenetic classification, they reduce the 'number of conclusions and predictions' they can make, and contradict Mayr's just quoted philosophy. The second defect is that Mayr's criterion of the 'best' classification is as subjective as it is flagrantly phenetic. There is no single most natural classification; the criterion of 'greatest number of conclusions and predictions' allows innumerable contradictory classifications. Cladism has entirely rejected phenetic criteria because they do not determine one 'best' kind of classification. It has moved instead to an objective, and justified philosophy. The philosophy will in some cases imply classifications less natural in the phenetic sense than the evolutionary and phenetic classifications; but the cladist regards that not as a defect but as an intention successfully fulfilled.

Cladism concentrates on shared derived characters because it is exclusively interested in the branching pattern of phylogeny. It does not attempt to represent all aspects of evolution; it does not attempt to represent differential rates of evolutionary divergence among groups: it only seeks to represent the hierarchy of recency of common ancestry. This is the hierarchy it is after, and this is the hierarchy it needs techniques to reveal. It could not concentrate so exclusively on shared derived characters if it were trying to represent phenetic similarity as well as branching. But it is not. It has philosophically rejected phenetic similarity. I therefore also disagree with Mayr[1] when he says 'it is evident that the cladist reveals great ambivalence in the

[1] Mayr (1974 [1976, p. 449]); 'reveals' is intended to reflect upon the cladist – reveals in himself – and not the ironical *double entendre* I should prefer – reveals in the other classificatory schools.

treatment of divergence'. He does nothing of the sort. Whatever its advocates may have said, the cladistic system contains no ambivalence whatsoever. Differential divergence has been scrapped. Of course that does not mean rates of evolution should not be studied: of course they should be. It is not part of classification, that is all. Like phenetic classification (p. 38 above), cladism should not be criticized for what it is not trying to do (cf. the exchange between Halstead 1978, Gardiner et al. 1979, and Halstead et al. 1979, and the references cited above, p. 32).

The well-justified philosophy of cladism gives to that school two methodic advantages relative to subjective phenetic systems. The first is that it facilitates methodic choice. Because cladists know what they wish their methods to do, when given a choice of several methods, they can tell which are better: a technique is as good as it is accurate at revealing the phylogenetic hierarchy. Secondly, it facilitates methodic improvement. Shared derived characters are an index of recency of common ancestry, but only an index. If someone suggests a new method it is easy, by the same criterion, to tell whether it is an improvement. In a system that lacks an objective philosophy, new methods will only proliferate; but in cladism, methodic innovation can be channelled into methodic improvement.

The distinction of ancestral from derived character states

The discussion so far as been purely theoretical. It could still turn out that Hennigian cladism is not practically possible. Let us now turn to its practical techniques. So far we have assumed we know whether a particular character state is ancestral or derived; now we must consider how this knowledge is acquired. It certainly cannot be discovered by direct observation, because the transitions between ancestral and derived states all took place in the past. Whatever techniques can be devised, they are going to be uncertain. Another point worth making at the outset, to forestall a common biological supposition (see, for example, the references of Schaeffer, Hecht, and Eldredge 1972, p. 38), is that the fossil record is not the only, nor the best, source of evidence: we shall come to it, but only after two other important sources of evidence.

We shall consider the cladistic techniques in two main stages. This section will discuss how to distinguish the ancestral from the derived conditions of single characters. The methods, as it happens, are all imperfect (although not useless), and there is a danger that they will make mistakes. If the methods were perfect, different characters

The techniques and justification of cladism

would all point to the same classification; but when a character state is wrongly identified (as ancestral for derived or vice versa), the character may well conflict with the evidence from other characters, and suggest a different classification. The next section accordingly will deal with the question of how the conflicting evidence of different characters may be reduced to a single classification. But we shall first concern ourselves with single characters: how can their ancestral and derived states be distinguished? Several techniques have been suggested, and they have been thoroughly reviewed (Crisci and Stuessy 1980; Stevens 1980); I am going to consider here only the three most important, to show how the distinction can be made, and to see how certain is the knowledge that they provide. The three, in the order we shall take them, are 'outgroup comparison', the embryological criterion, and the palaeontological criterion. I shall, for the most part, be assuming that the characters under analysis are homologies; some techniques (such as those discussed in Chapter 2) to distinguish homologies from analogies should have already been applied. Those techniques, as we have seen, were uncertain; and convergence will provide a second reason why different characters may suggest contradictory classifications, which is the problem to be discussed in the next section.

Outgroup comparison

We can see outgroup comparison in its simplest form in the case of one character in a pair of species, one of which has the character state a, the other a' (Figure 4.7a). Which state is ancestral, and which derived? The answer, by outgroup comparison, is obtained by examining a related species, which is called the outgroup. The state of the character in the outgroup is ruled to be ancestral. If the outgroup has a, a' is the derived state (Figure 4.7b). If species 1 and 2 were (for example) a reptile and a fish, character a' was a cleidoic egg, and a a non-cleidoic egg, we could take an outgroup such as an echinoderm to determine that the cleidoic egg of the reptile was the derived state. The method relies on the principle of parsimony, according to which the number of evolutionary events has been the minimum possible, given that the species have the character states they do: shared characters therefore are more likely to be due to common ancestry than to convergence. The principle is used in some form or other in most methods of reconstructing evolution, and can be justified by assuming that evolutionary change is relatively improbable (Felsenstein 1983a, b); it has erroneous justifications too, some of which we shall notice later. The number of evolutionary events in Figure 4.7 is indeed the minimum possible: if the common ancestor of 1 and 2 had been taken to be a', then there would have

Evolution and Classification

Species	1	2
Character state	a	a'

Species	1	2	Outgroup
Character state	a	a'	a

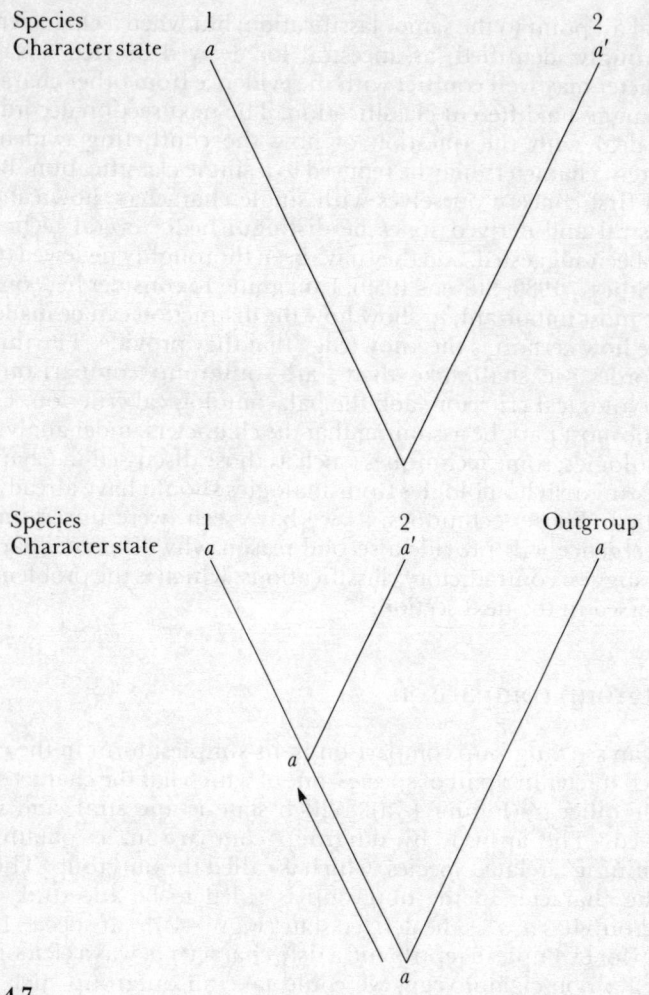

Figure 4.7
(a) Two species, 1 and 2, each with a different state (*a* and *a'*) of the same character (*A*). The cladistic problem is, which state is ancestral and which derived? (b) The problem solved by outgroup comparison. The state of the character is examined in a related species (the outgroup). Its state in the outgroup is taken to be ancestral in the pair of species 1 and 2.

been at least two events (the evolution of *a'* between the outgroup and the common ancestor and the re-evolution of *a* in species 2); if the common ancestor had been taken to have some other state, such as *a''*, even more events could be counted.

Exactly which species should we take as the outgroup? One reasonable rule would be to pick any related species. As long as they

The techniques and justification of cladism

all have the same state, they will all give the same answer, and it will not matter which is used. If, for example, the character states were the presence and absence of a backbone, and the two species were a fish and an echinoderm, then any outgroup will imply that its absence is ancestral. But what if different outgroups suggest different ancestral states (Figure 4.8)? Whatever the method, the conclusion is now going to be less certain. With characters like those in Figure 4.8 the assumption of outgroup comparison, that convergence is relatively rare, is dubious. But the assumption can still be applied, and leads to clear rules on how to proceed (Maddison, Donoghue, and Maddison 1984). In the case of Figure 4.8 the estimated evolutionary change will be minimized by taking the most closely related species as the outgroup.

There are two main lines of criticism of outgroup comparison. One challenges its assumption that convergence is relatively rare. If it were false, if convergence were more probable than shared ancestry, the principle of parsimony would fail, and outgroup comparison would not distinguish ancestral from derived character states. If a was convergent in species 1 and the outgroup (in Figure 4.7), then it would be derived in species 1, not ancestral as outgroup comparison asserts. Parsimonious arguments, such as that convergence is minimal, crop up in several different places in cladism, and they can always be accompanied by the criticism that evolution is not in fact parsimonious (Cain 1967, 1982; Friday 1982). The critic should be

Species	1	2	3	4	5	6
Character state	a	a'	a	a'	a'	a

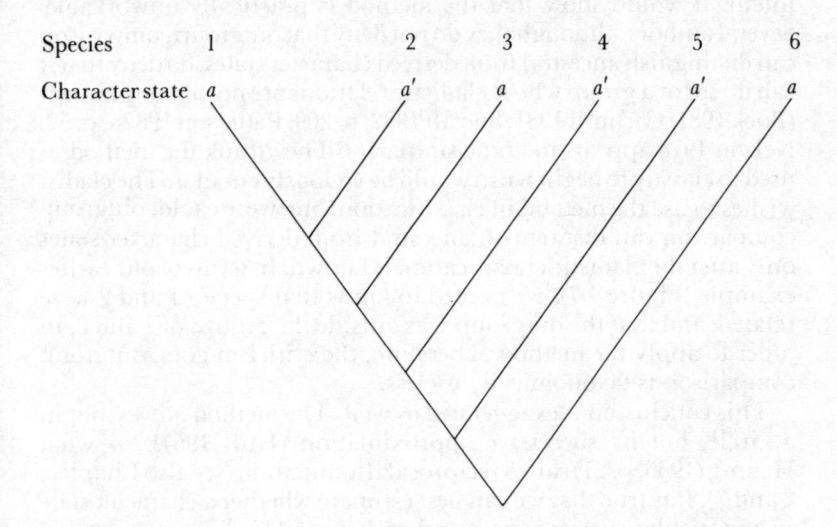

Figure 4.8
Subversion of outgroup comparison. Different candidate outgroups of species 1 and 2 have different states for the character in question. Outgroup comparison is here difficult, but not impossible, to apply.

Evolution and Classification

careful about what parsimony means. If, for instance, it meant that convergence does not take place (as it is often said to), it would obviously be wrong, for we know that convergence does in fact occur. But the principle makes no such claim. It does not say that convergence is non-existent, nor even that it is rare (Farris 1983): it only says it is *relatively* rare. Whenever it is criticized, therefore, the same general reply may be made. Although the principle is imperfect, it is not absurd. Its proper justification is not known to be false, and is reasonable. The principle is probably nearer the truth than any other practical assumption that can be made. The criticism can only be destructive: it offers no alternative method. Moreover, it can easily be exaggerated. If the principle of parsimony were completely unreliable, we should have no knowledge of evolution at all: we should not know that humans are any more closely related to apes than to worms. The principle may not be perfect, but it has some validity. It can be used to reconstruct phylogeny, but we should not have excessive confidence in the conclusions. It is not a mistake to criticize parsimonious arguments, it is only a mistake to exaggerate the criticism into total rejection. Sensible critics may remark that the principle of parsimony is an assumption in need of improvement; but they should not reject it entirely. If the principle were rejected, all our knowledge of evolution (and probably of much else besides) would come tumbling down. Few biologists would accept that result.

The second criticism is different in form but equally destructive in intent. It would show that the method is practically unworkable. Several authors, although they do not deny that outgroup comparison can distinguish ancestral from derived character states, do deny that it can do so for a group whose cladistic relations are not already known (Bock 1981; Cartmill 1981; Sneath 1982, p. 209; Patterson 1982b, p. 52; Nelson 1978 appears to argue similarly). They think the method, if used to classify to begin with, would be viciously circular. The cladist wishes to use the method in classification; but, we are told, outgroup comparison can distinguish ancestral from derived character states only after the cladistic classification is known. In terms of our earlier example (Figure 4.7), we needed to know that species 1 and 2 were related, and that the outgroup was outside the groups of 1 and 2, in order to apply the method. Therefore, the criticism goes, outgroup comparison is taxonomically useless.

This conclusion is exaggerated as well. The method moves, not in a circle, but by successive approximation (Hull 1967), or what Hennig (1966, p. 21) called reciprocal illumination (see also Chapters 2 and 7). It is true that we can best estimate whether a character state in a particular species is ancestral or derived if we know the species' genealogy. But if the phylogeny is imperfectly known, an estimate can still be made. The estimate will still be better than to suppose that the character states are ancestral and derived in random proportions:

The techniques and justification of cladism

it will only be less good; it will not be utterly misguided. It is perfectly possible to form some tentative, and crude, idea of the phylogenetic relations of species, and then to apply outgroup comparison. Suppose (Figure 4.9), for example, that the relations of five species (numbered 1 to 5) are unknown, but that some other species is thought to be less related to them than any are to each other. That other species can act as an outgroup. We must first find out the states of a character in all six species. It may be that it has two states, and that species 2, 3, and 4 share the state of the outgroup, and species 1 and 5 share the derived state. Then we can modify the cladogram, to make species 1 and 5 a sister group (Figure 4.9c). This procedure can be continued with further characters. After sufficient facts had been considered, it might prove necessary to modify the initial idea that species 6 was an outgroup: if, for instance, one character after another suggested that species 1, 5, and 6 formed a group, then the whole cladogram would have to be modified. If our initial guess turns out to imply the recurrent evolution and loss of a backbone on dozens of occasions, then it could be changed to one which did not. As each new character is studied the relations of the species are re-assessed. When the assessment changes, all the previously studied characters would have to be re-assessed in turn. But if there is a cladistic classification, then after enough characters have been studied outgroup comparison should eventually reveal it. The method is least effective at the beginning of classification, when little or nothing is known about the relations of the species. As knowledge increases, the method becomes more and more powerful. For instance, once we know that

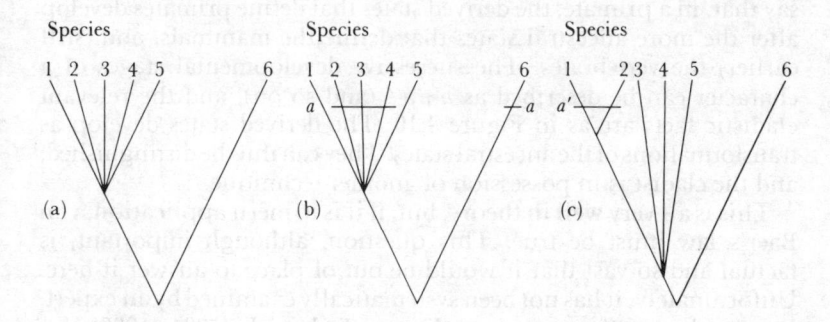

Figure 4.9
The method of successive approximation. (a) The cladistic relations of five species (1 to 5) are not known. The outgroup is tentatively thought to be less related to them than they are to each other. (b) The states of a character A are examined in all six species. The state in the outgroup is taken to be ancestral. Species 1 and 5 here share the derived state, which is evidence that they are cladistic sister groups. (c) The evidence is used to modify the cladogram.

Evolution and Classification

the vertebrates are a separate group from invertebrates, we can use invertebrates as outgroups to sort out the relations within the vertebrate groups. The method is therefore highly important in classification. The kind of circularity it involves is not vicious. And even if it were completely circular (which it is not) - that is, if it could only distinguish ancestral and derived character states after a cladogram had been formed - it would still be possible to use it to check the classifications made by other cladistic techniques. The sensible cladist employs the technique.

The embryological criterion

Now for the embryological criterion. Embryological evidence is useful in classification because the courses of development of related species are similar, just as are there adult phenotypes. The embryological distinction of ancestral from derived character states, however, requires something more precise than that. To be exact, it requires the truth of the first of von Baer's laws: 'The general features of a large group of animals appear earlier in the embryo than the special features' (von Baer 1828, p. 224, put into English by Gould 1977a, p. 56). In the development of a primate, for example (if the law is correct), the developmental stages common to all vertebrates develop before those common only to mammals, which in turn develop before those common only to primates. This law can be put to cladistic use, after a simple re-interpretation. The special characters must be interpreted as derived character states, and the general characters as ancestral states. With this interpretation, we can say that, in a primate, the derived states that define primates develop after the more ancestral states that define the mammals, and (still earlier) the vertebrates. The successive developmental stages of a character can be described as a', a'' (and so on), and the relevant cladistic facts are as in Figure 4.10. The derived states develop as transformations of the ancestral states. They can thus be distinguished; and the cladist is in possession of another technique.

This is all very well in theory, but, if it is to merit application, von Baer's law must be true. This question, although important, is factual and so vast that it would be out of place to answer it here. Unfortunately, it has not been systematically examined by an expert. Experts have offered their opinions: Sedgwick (1894, 1909), for instance, provided some counter-examples and said that the law was false; Weismann (1882, pp. 390-554) compared larval and adult classifications of Lepidoptera, and found some cases in which larvae differed more than adults (see especially pp. 435-8), as well as some in which the larvae were more similar (as von Baer's law would predict); Weismann's study was followed up in a minor way by van Emden

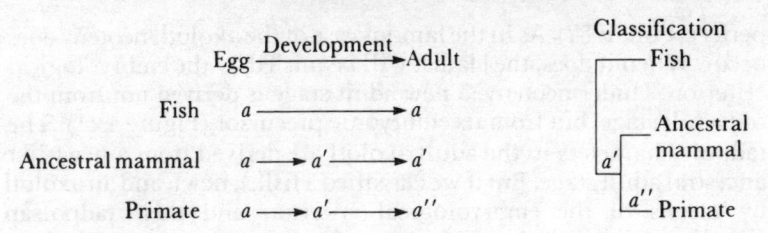

	Development	Classification
Egg ──────────────→ Adult	┌── Fish
Fish	a ─────────→ a	
Ancestral mammal	a ──→ a' ─────→ a'	a' ┤ ┌── Ancestral mammal
Primate	a ──→ a' ─────→ a''	└a''┤ └── Primate

Figure 4.10
The embryological criterion. Successive developmental stages (a, a', a'') are taken to be successively derived states of a character. The derived states can be used to form a cladogram (at right).

(1929, and see also de Beer 1940); further counter-examples are mentioned by Rieppel (1980) and Patterson (1983, p. 25). But I have found no general review. The most likely source, Professor Gould (1977a, p. 232), when he arrives at the point in his great work where he should examine the law, promptly retreats down his favourite philosophical bolt-hole, declaring that 'empirical tabulations will not solve the problem; there are simply too many cases to count and prior attitudes always dictate the selection'. But that will not do, for even if there is too much material for an industrious reviewer (which I doubt), a random taxonomic sample could be used, or even one group could be systematically investigated as Weismann did for the Lepidoptera.

In the absence of any authority we shall have to content ourselves with some more general observations. The law relies on what Gould (1977a) calls 'terminal addition', the principle that new characters in evolution are added on to the end of the existing ontogeny, rather than being intruded earlier on. If the principle of terminal addition is true, so too will be von Baer's law. Cases of evolution by non-terminal addition are known: they fall into the general categories of neoteny and paedomorphosis, but there has been no systematic study of whether they are the rule or the exception. Some authors (de Beer 1931; Hardy 1954) give the impression that terminal addition is, at most, no more usual that neoteny; others (cited by Gould 1977a, p. 232) suspect that terminal addition is normal. We can only conclude that terminal addition is at least sufficiently common for von Baer's law to possess some use for the cladist. It is difficult to settle the factual question when we do not know what to expect theoretically. There are arguments to explain why evolution should tend to be by non-terminal (Hardy 1954) or by terminal (Gould 1977a, pp. 231–4; Ridley 1986) addition, but we do not know if there is any truth in them.

Von Baer's law probably is true enough to be usable, but it is not

Evolution and Classification

perfect (Fink 1982). As in the famous case of the axolotl, neoteny does occur. When it does, the cladist will be misled by the embryological criterion. Under neoteny, a new adult stage is derived not from the old adult stage, but from its embryonic precursor (Figure 4.11). The tadpole characters in the adult axolotl are derived from a newt-like ancestral adult stage. But if we classified a fish, a newt, and an axolotl by means of the embryological criterion and their tadpolean characters, the axolotl would turn out as the sister species of the fish. In fact it is the sister of the newt.

We do not know how often, or in what circumstances, the embryological criterion is valid. Its truth is proportional to that of the principle of terminal addition: but we do not know how frequent terminal addition is in nature; and we do know that, at least sometimes, the principle is wrong. We must, in the end, deliver the same judgement on the embryological criterion as we gave for outgroup comparison. It is imperfect; it is better than nothing; further investigation may improve it. Its particular imperfection, however, differs from that of outgroup comparison. Outgroup comparison requires the absence of evolutionary convergence; but the embryological criterion requires all evolutionary additions to be ontogenetically terminal. The relative merits of outgroup comparison and the embryological criterion are at present controversial (Stevens 1983).

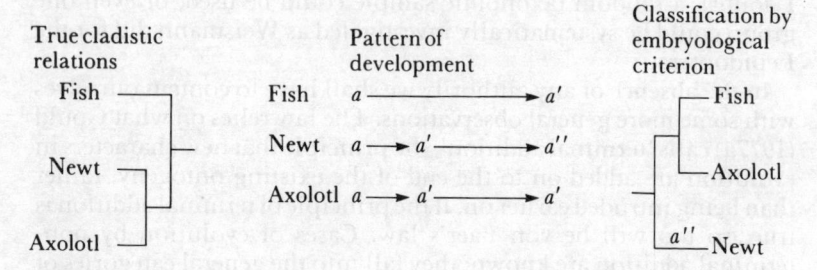

True cladistic relations	Pattern of development	Classification by embryological criterion
Fish	Fish $a \longrightarrow a'$	Fish
Newt	Newt $a \longrightarrow a' \longrightarrow a''$	Axolotl
Axolotl	Axolotl $a \longrightarrow a' \longrightarrow a'$	a'' Newt

Figure 4.11
Neotenic subversion of the embryological criterion. The true evolutionary relations are at left. The axolotl has neotenously evolved, from a newt-like ancestor, to possess tadpole-like features. The tadpole-like features, in the adult, are derived, but they look ancestral. When they are taken to be ancestral, the embryological criterion lies (cladogram at right).

The palaeontological criterion

Let us turn finally to palaeontology. Ancestral states must have evolved before derived ones. It might be supposed, therefore, that if we looked at earlier and earlier fossils, we should see successively

The techniques and justification of cladism

more ancestral character states. The cladistic distinction could then be made by direct observation. Unfortunately the fossil record is not complete. Many characters leave no fossil record at all, and besides, for those characters that are preserved in fossils, the record is fragmentary. An ancestral state must indeed precede its derived form in evolution, but that does not mean that it must precede it in the fossil record. If the ancestral form is not preserved in earlier strata, but is preserved in later strata, and if the derived state is preserved in earlier rocks, then the palaeontological criterion will reverse the truth (Figure 4.12). The palaeontological criterion, therefore, may lie. But that is not to say that it will be wrong every time. If the earliest ancestral forms are preserved, the fossil record will tell the truth. In general, if the gaps in the fossil record are distributed randomly with respect to ancestral and derived states, the fossil record will be right more often than not. Experts differ as to the merit of the palaeontological criterion; some think it good (Paul 1982), others think it useless (Schaeffer, Hecht, and Eldredge 1972; Patterson 1981; Janvier 1984). Certainly it can work only with a limited range of groups and characters. I do not think it should be entirely rejected, but it is probably practically less valuable than the other criteria.

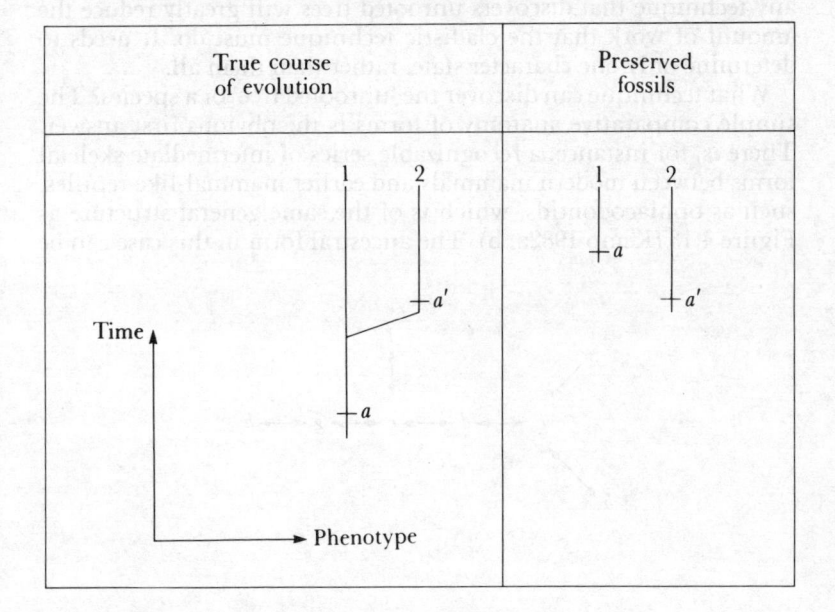

Figure 4.12
Subversion of palaeontological criterion by the incompleteness of the fossil record. The true pattern of evolution is on the left: a is the ancestral state, a' derived. Selective loss of the ancestral species makes a' appear ancestral, and a derived.

Evolution and Classification

Unrooted trees

Each of the three main cladistic techniques can distinguish ancestral and derived character states in pairs of species that differ in their character states. In practice, of course, there are thousands of species to be classified, and if the techniques in the forms described were the limit of our repertory, we should have to apply the techniques in turn to small groups of species, to work out gradually the cladistic relations of the whole. However, there are other techniques which, although they do not determine the polarity of character states, do order species into evolutionary series such as Figure 4.13, which shows the kind of arrangement often called an 'unrooted tree'. By itself an unrooted tree does not determine a classification, because it allows as many classifications as there are species, according to which species is ancestral (Figure 4.14). But if the most ancestral state (the 'root') can be found, all the polarities of all the other character states will be implied. Although an unrooted tree does not specify a classification, it places strong constraints on the possibilities. One of the cladistic techniques such as outgroup comparison, the embryological criterion, or palaeontology, will be needed to root the tree, but any technique that discovers unrooted trees will greatly reduce the amount of work that the cladistic technique must do. It needs to determine only one character state, rather than them all.

What technique can discover the 'unrooted tree' of a species? The simple comparative anatomy of forms is the obvious first answer. There is, for instance, a recognizable series of intermediate skeletal forms between modern mammals and earlier mammal-like reptiles, such as ophiacodontids, which is of the same general structure as Figure 4.13 (Kemp 1982a, b). The ancestral form in this case can be

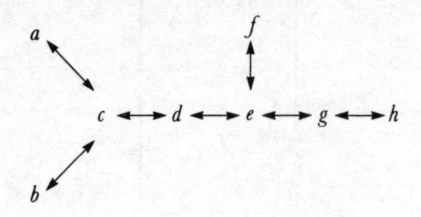

Figure 4.13
An unrooted tree. The direction of the phylogenetic relations among species is uncertain. Among species *a, c,* and *d,* for example, the pattern of evolution may have been $a \rightarrow c \rightarrow d$, or $a \leftarrow c \leftarrow d$, or $a \leftarrow c \rightarrow d$, but not $a \rightarrow d \rightarrow c$ or $c \rightarrow a \rightarrow d$.

The techniques and justification of cladism

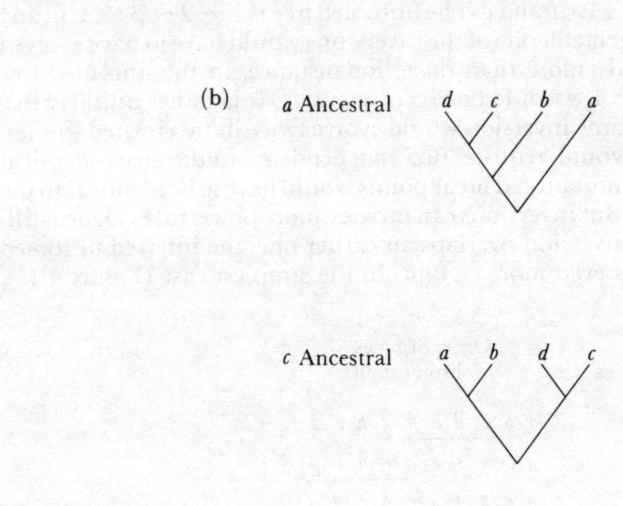

(a) $a \leftrightarrow b \leftrightarrow c \leftrightarrow d$

(b) a Ancestral $d\ c\ b\ a$

c Ancestral $a\ b\ d\ c$

Figure 4.14
Unrooted trees do not determine cladistic classifications. The cladistic classification of the simple unrooted tree (a) depends (b) on which species is ancestral. If, for example, a is ancestral, the classification differs from that if c is ancestral.

determined palaeontologically or by outgroup comparison. The species in question are fossils, distributed in time starting about 315 million years ago. The palaeontological criterion (uncontroversially) suggests that the therapsid and then cynodont conditions are ancestral to those of modern mammals. Outgroup comparison with the other reptiles and amphibians would suggest the same root to the tree. Only the embryological criterion could not be applied to this fossil material.

The phylogeny of Hawaiian *Drosophila* has been worked out by an exceptionally powerful version of the same class of technique. The characters in this case are chromosomal inversions (Carson and Kaneshiro 1976; Carson and Yoon 1982; Carson 1983; see Williamson 1981). The two ends of chromosomal inversions can be located with exceptional accuracy in the giant polytene forms of the chromosomes that can be found in the larval salivary glands. Inversions are

Evolution and Classification

described relative to an arbitrarily selected 'standard' chromosome, and can be symbolized by single letters. They reveal unrooted trees in two ways. The first is the same as for any other character. Suppose, for example, there are three separate inversions, *a, b, c,* and species 1 has none of them (it is the 'standard'), species 2 has *a* only, species 3 *a* and *b*, and species 4 *a, b* and *c*. The unrooted tree is 1 ↔ 2 ↔ 3 ↔ 4: in any other arrangement, one of the inversions would have to have evolved independently more than once; for instance, in the unrooted tree 2 ↔ 1 ↔ 3 ↔ 4, *a* would have evolved twice. It is rather unlikely that exactly the same inversion would evolve twice in two related species, because it would require *two* independent, undirected, identical events: both mutational breakpoints would have to be identical in the two species. But inversions can provide more powerful evidence still. If a second inversion overlaps an earlier one, the inferred unrooted tree becomes even more certain. In the simplest case (Figure 4.15),

(a) Species   Order of genes on chromosome

     1   *a b c d e f g h i j*

     2   *a b g f e d c h i j*

     3   *a b g f e i h c d j*

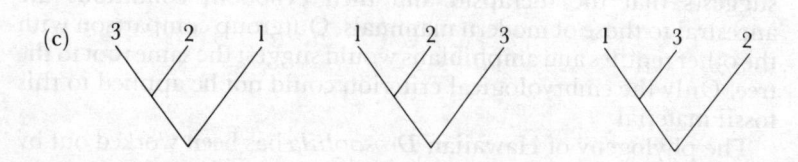

(b)   1 ↔ 2 ↔ 3

(c)   3   2   1   1   2   3   1   3   2

Figure 4.15
Inference of unrooted trees by means of chromosomal inversions. (a) Gene sequences on the chromosomes of three species, 1, 2, and 3. (b) The unrooted tree, which allows (c) three possible classifications according to which species has the ancestral chromosome.

convergence requires four independent identical breakpoints; and as the number of overlapping inversions and species increases, the chances of convergence become vanishingly small. It becomes most unlikely that, if the inversion is mapped accurately, convergence will be overlooked. The phylogenetic relations suggested by different sets of inversions therefore should not differ. One does not expect character conflict. The phylogenetic inference has exceptional certainty.

The possibility of using chromosomal inversions to infer phylogenetic relations was first put into practice by Wasserman (1954, summarized in Wasserman 1963), who applied it to the *repleta* group. Its most impressive application, however, has been by Carson and his colleagues (starting with Carson, Clayton, and Stalker 1967), to the 'picture-wing' groups of Hawaiian drosophilid flies. There are believed to be about 700 species of *Drosophila* in the Hawaiian archipelago, of which about 110 are in the picture-wing group. The most recent paper by Carson (1983) gives the phylogeny of 103 of these, estimated with 214 chromosomal inversions (Figure 4.16). How can the tree be rooted? Here, there are two classes of evidence, outgroup comparison and geology. A pair of sibling species, *Drosophila primaeva* and *D. attigua,* have the chromosomal pattern of what are thought to be closely related continental forms, from which the Hawaiian *Drosophila* originated: outgroup comparison therefore places them at the root of the tree (Stalker 1972; Williamson 1981, p. 179). The geological history of the islands is also suggestive. Kauai is the oldest island, Hawaii itself the youngest; the ancestor of the group probably first colonized Kauai. We should expect it to remain there, because almost every species is confined to a single island. And *Drosophila primaeva* and *attigua* do inhabit only Kuaia. Both lines of evidence therefore suggest they are the root of the tree, the ancestor of the Hawaiian picture-wing *Drosophila.* Once that is determined, all the other cladistic relations are as well. One only need find the root of an unrooted tree and the whole cladistic classification is implied. The kind of reasoning could be thought of as a successive outgroup comparison through the whole tree: once the root species is known, it can act as outgroup to the next groups of species (*ornata* and species nos 4–16 in Figure 4.16); but this is an unnecessary interpretation. Once the one species at the root has (by outgroup comparison) been found, the unrooted tree implies the whole phylogeny, the whole cladistic classification.

No other study of a comparably large group of species can stand comparison with the work on Hawaiian *Drosophila.* The fact that none of the 214 inversions suggest different phylogenies (H. L. Carson, personal communication) gives some indication of the power of the technique. One might hope that as the same method was applied to other plants and animals, it would yield similar results.

Evolution and Classification

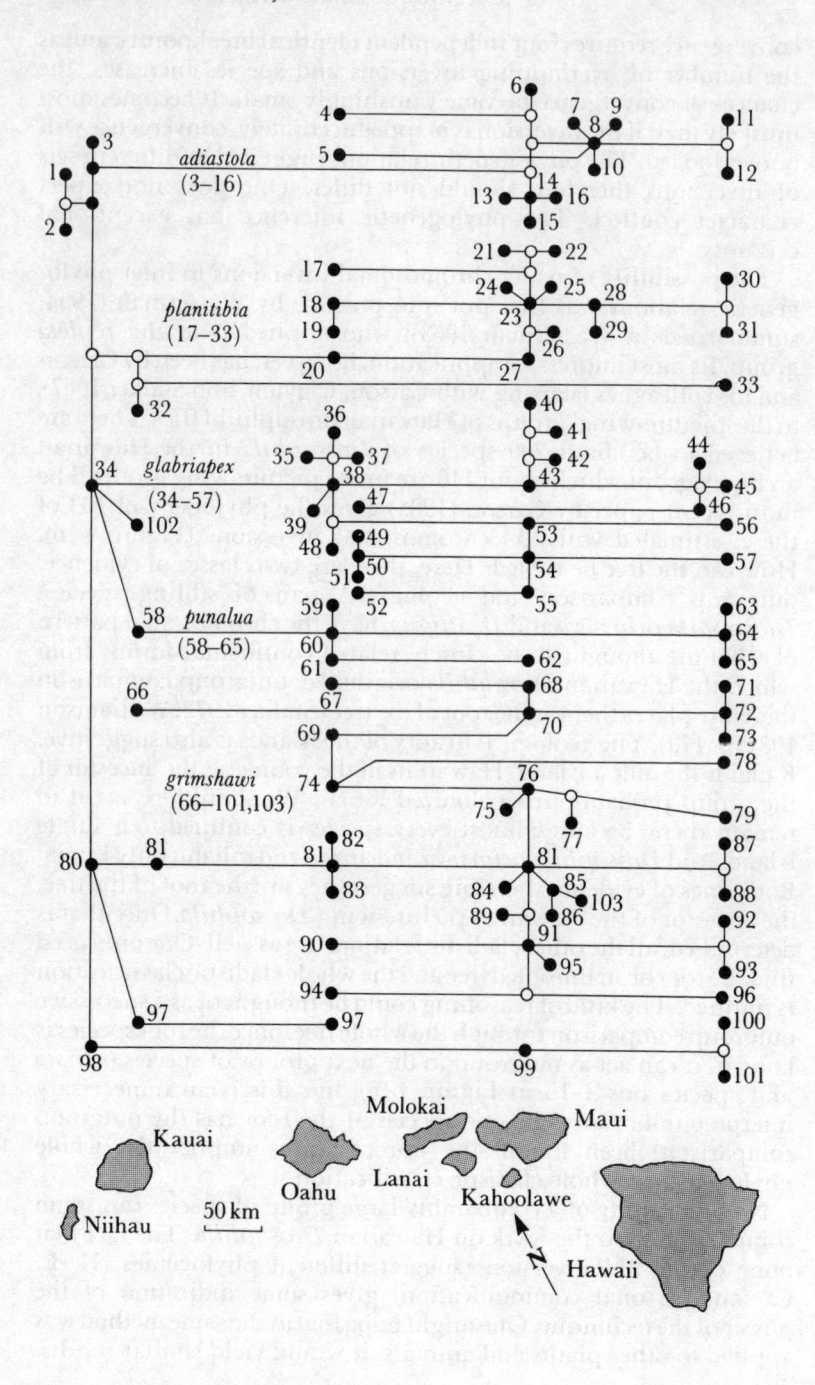

adiastola
(3–16)

planitibia
(17–33)

glabriapex
(34–57)

punalua
(58–65)

grimshawi
(66–101,103)

Kauai Oahu Molokai Lanai Maui Kahoolawe Hawaii

50 km

N

74

Unfortunately, however, the method may have limited scope. The banding patterns of *Drosophila* chromosomes can be determined with exceptional accuracy in their giant polytene forms. If the ends of the inversions are located less accurately, it becomes easier to confuse different inversions. The same (as it appears) inversion may then seem to have evolved convergently in different groups. Different inversions will then contradict each other, and chromosomal inversions have the same difficulties as other characters. There are indeed difficulties with contradictory inversions among the species

Figure 4.16
Phylogeny of 103 species of Hawaiian *Drosophila* of the 'picture-wing' group. Chromosomal patterns fixed the unrooted tree, which was rooted by outgroup comparison and the geochronology of the islands. Some details of the tree are dictated by biogeography rather than chromosomal inversions. Species shown as ancestors may be relatives of an unknown ancestor: it takes extra steps to depict the phylogeny without any species being shown as the ancestor of any other.

1 *primaeva*	2 *attigua*	3 *ornata*	4 *neogrimshawi*
5 *touchardiae*	6 *clavisetae*	7 *cilifera*	8 *adiastola*
9 *peniculipedis*	10 *spectabilis*	11 *setosimentum*	12 *ochrobasis*
13 *hamifera*	14 *paenehamifera*	15 *truncipenna*	16 *varipennis*
17 *nigribasis*	18 *oahuensis*	19 *hemipeza*	20 *substenoptera*
21 *melanocephala*	22 *ingens*	23 *neoperkinsi*	24 *cyrtoloma*
25 *hanaulae*	26 *neopicta*	27 *obscuripes*	28 *planitibia*
29 *differens*	30 *silvestris*	31 *heteroneura*	32 *picticornis*
33 *setosifrons*	34 *glabriapex*	35 *inedita*	36 *distinguenda*
37 *divaricata*	38 *pilimana*	39 *aglaia*	40 *discreta*
41 *lineosetae*	42 *fasciculisetae*	43 *vesciseta*	44 *alsophila*
45 *conspicua*	46 *assita*	47 *montgomeryi*	48 *hexachaetae*
49 *spaniothrix*	50 *tarphytrichia*	51 *psilophallus*	52 *gymnophallus*
53 *virgulata*	54 *odontophallus*	55 *liophallus*	56 *digressa*
57 *macrothrix*	58 *ocellata*	59 *uniseriata*	60 *punalua*
61 *paucicilia*	62 *prostopalpis*	63 *basisetae*	64 *paucipuncta*
65 *prolaticilia*	66 *musaphilia*	67 *turbata*	68 *recticilia*
69 *gradata*	70 *gymnobasis*	71 *heedi*	72 *hawaiiensis*
73 *silvarentis*	74 *flexipes*	75 *lasiopoda*	76 *villitibia*
77 *hirtipalpus*	78 *formella*	79 *psilotarsalis*	80 *villosipedis*
81 *grimshawi*	82 *atrimentum*	83 *obatai*	84 *sodomae*
85 *disjuncta*	86 *bostrycha*	87 *sproati*	88 *pullipes*
89 *orphnopez*	90 *sobrina*	91 *balioptera*	92 *murphyi*
93 *engyochracea*	94 *reynoldsiae*	95 *orthofascia*	96 *ciliaticrus*
97 *crucigera*	98 *sejuncta*	99 *limitata*	100 *ochracea*
101 *claytonae*	102 *micromyia*	103 *affinidisjuncta*	

Slightly updated Williamson (1981), using information in Carson (1983) and a private communication from Williamson; Williamson worked from Carson and Kaneshiro (1976).

Evolution and Classification

of great apes (de Grouchy, Turleau, and Finaz 1978): the method is not exceptionally powerful here.

Techniques that discover unrooted trees are a great aid in cladism, but they do not by themselves fix the classification. In cladism, they are neither necessary nor sufficient; but that does not mean they should not be used. They should be used whenever they can. To classify species the cladist will still rely finally on techniques that distinguish absolutely which character states are ancestral and which are derived; on, in particular, outgroup comparison, the embryological, and palaeontological, criteria. An unrooted tree, however, places a strong constraint on the form of the classification; it is an independent source of evidence against which the other techniques can (to some extent) be checked. They should be so checked. None of the cladistic methods is perfect. Indeed their imperfections are only too obvious. If we concentrate on the circularity of outgroup comparison, the theoretical pipe-dreams and factual fragility of the embryological criterion, the incompleteness of the fossil record, we may well wish to conclude (like Pantin's student) that outgroup comparison is mute, embryology meaningless, and palaeontology lies, before moving on to some calling less arduous than cladism. But another less-pessimistic conclusion is possible. All the methods do possess some validity, and if we take them together and challenge the states of characters with each of them in turn, we may arrive at some practically reliable results. Each of the methods may make mistakes, and the more they are checked against each other, the more likely it is that their mistakes will be uncovered. It is important to realize that, because each method, taken by itself, is prone to error, we should use as many lines of evidence as possible.

Many cladists do not realize this. Patterson (1981), for example, has subjected the palaeontological criterion to a severe analysis, in which he shows that it has never overruled any of the other criteria. He thus proved that it is dispensible; but he then concluded that it should be dispensed with. We shall meet an analogous argument, by Cracraft, in Chapter 8: he dispatched the functional criterion after arguing that it only imperfectly overlaps other techniques. But, given the true nature of the techniques, these arguments are mistaken. All the methods are imperfect; taken individually, they are all dispensible; but the best method is still to use as many of them as possible. Some cladists have fired a related argument against evolutionary taxonomy, rather than against other cladistic factions. Hull (1979, 1980) has drawn attention to the recurrent cladistic habit of taking together the uncertain methods of the cladist and the uncertain methods of the evolutionary taxonomist, criticizing the imperfections of the latter, ignoring those of the former, and concluding that evolutionary taxonomy is unworkable.

'Never trust to the principle of exclusion!' When all the alternatives are uncertain, the conclusions of this method depend crucially on the order in which the alternatives are considered. If palaeontology is taken first, it can be made to appear dispensible: but the same could be done with any of the other techniques. If we rely on the method of exclusion, the cladist's repertory of techniques will end up depending, not on the relative merits of the techniques, but on the starting point of certain cladistic discussions of the techniques. That is an absurdity to be avoided.

'Hennig's dilemma': the disagreement of characters

In theory, all the shared derived characters should fall into the same groups, which (as Figure 4.5 illustrates) unambiguously make up the cladistic classification. In practice, however, the character states thought to be derived may be distributed in contradictory groupings, for two reasons. If ancestral character states are mistaken for derived ones, they may contradict each other and any correctly assessed derived characters. Ancestral characters can be retained by many kinds of non-cladistic groups, or by cladistic groups, depending on the pattern of evolution (Figure 4.4). The other reason is unrecognized convergence. Whereas ancestral characters merely may contradict the cladistic classification, convergent ones nearly always do. The only case in which they do not is when a derived character secondarily converges back on to its ancestral state; then if that ancestral state happens to fit the cladistic classification, two wrongs will make a right, and the double mistake will not create any difficulty. In other cases, unrecognized convergence will cause different apparently derived characters to fall into contradictory groups. What are really different characters in the different species will appear to be the same derived character. (I should perhaps reiterate how I am using the term 'character'. 'Shared derived character' could mean two things. It could refer only to the descendants of one evolutionary change: these are the shared derived characters of Figure 4.4 and used to define cladistic groups. However, if a character converged onto an identical form after more than one evolutionary change in separate evolutionary lineages, we might also call it a shared derived character. In either case the problem is to distinguish the former kind of shared derived characters from the latter kind; but we must decide whether to use the term to refer broadly both to the apparent and real cases, or narrowly, only to the real ones. I shall use the narrower meaning. A shared derived character hereafter is one that is shared by common descent from a single evolutionary change. In other words, 'character' here refers to a unique section of the evolutionary history of a property of an organism; it is a theoretically defined term, the instances of which have to be inferred: it could have been defined operationally to refer to things that look the same. My choice of theoretical rather than operational meaning does not alter the problem to be solved.)

If convergence is recognized it presents no problem. The convergent characters will be appropriately identified as the different character that, in evolutionary terms, they are; they are as easy to deal with as any other correctly identified character. Different correctly identified shared derived characters, in our narrow meaning, *must* agree. They must define the same hierarchy of species, because all the evolutionary changes have been forced through the same true phylogeny. But often the true shared ancestral and derived characters will be wrongly inferred. The different apparently shared derived characters may then not agree, and it is not so easy to produce a cladogram from the evidence. The Arthropoda, for example, provide a case of notorious uncertainty. Ignoring many other disagreements among characters, the insects, *Peripatus*, and crustaceans seem to share mandibles and the insects and arachnids share tracheae; insects, crustaceans, and arachnids, but not *Peripatus*, have compound eyes. Some of what are thought to be shared derived character states are clearly convergent. How should we proceed? The obvious next move is to study the conflicting characters further in an attempt to discover which are convergent. Any of the criteria discussed in Chapter 2 for distinguishing homologies from analogies could be applied, to show that what looks like a shared derived character (and therefore homologous) is in fact two different characters (and therefore analogous). Thus Manton (1977) studied the mouthparts of all the arthropod groups and concluded that the similarities of insectan and crustacean mandibles are only superficial and convergent. Her conclusion is controversial; but let that pass here. The point is, further study of disagreements among characters may remove the disagreement, and if it does the problem is removed.

But what if further study draws a blank? This is the problem Felsenstein (1982) called 'Hennig's dilemma'. There are several possible courses of action. One is to give up phylogenetic classification and fall back on a subjective phenetic system. The other is to stay within the phylogenetic system, and either hang fire until new evidence becomes available, or use one of the statistical methods of phylogenetic inference from conflicting evidence (reviewed by Felsenstein 1982, 1983a,b). The statistical methods have grown up as large amounts of molecular evidence have accumulated, to which they are particularly appropriate, both because molecules provide many comparable characters and also because the other option is inapplicable: no amount of further study is likely to remove character disagreements. Moreover, disagreements among molecular characters, such as the amino acids of a protein, are very common. Figure 4.17 illustrates the variable part of the sequence of cytochrome c in 21 vertebrate species. There are many conflicting similarities: take amino acid sites 3 and 12, for example. Site 3 divides the 21 species into two groups: the chicken, turkey, and pigeon on the one hand share isoleucine, and all the others, which share valine. Site 12 also divides the 21 species into two groups: but this time into a group of man, chimp, and rattlesnake (which share methionine) and all the rest (which share glycine). The amino acid contradictions have two familiar sources, ancestry and convergence. The grouping defined by site 3 disagrees with all other 103 sites, because of shared ancestral similarity. At site 3, valine is probably the ancestral state, and the group of 18 species except chicken, turkey, and pigeon (man, chimp ... tuna, dogfish) is defined by a shared ancestral character. If the ancestral states are correctly identified, cladism removes the disagreement due to them, by ignoring them. But the other source of disagreement, that due to convergence, remains. The group of rattlesnake, chimp, and man defined by site 12 again disagrees with the other sites; and it is probably polyphyletic. Cladism removes the disagreement due to different shared ancestral amino acids at different sites, but is left with the disagreement due to convergence. For molecules, disagreements probably will not be removed by further study: we cannot expect further work to reveal that the methionine of the rattlesnake is really some other amino acid. Different sites suggest different groupings: one of them must be correct, but how are we to recognize it?

Felsenstein (1982) has reviewed several statistical methods. The most familiar and most used is the minimum evolution tree. The tree of minimum evolution is the one that implies the smallest number of evolutionary events, the least convergence. The procedure, given the distribution of characters among species, is to count the minimum number of evolutionary events implied by all the possible phylogenies of the species. The phylogeny is chosen that requires the fewest events; it is in a sense the most parsimonious phylogeny. From the evidence of Figure 4.17, the cladogram of maximal parsimony would not put the rattlesnake, chimp, and man together merely because they are similar at one or two sites; at all the rest they are not: if the similarity of chimp and rattlesnake is inferred to be convergent, only two amino acids sites would have converged. The alternative, to suppose that the rattlesnake and chimp share a more recent common ancestor than do (say) chimp and dog, would imply many dozens of convergent changes. The most parsimonious phylogeny is the one that implies the least possible convergence, given the status of all the species. As a method, it has two difficulties. One is practical: it is difficult (or even practically impossible – Felsenstein 1982, p. 387) to

Evolution and Classification

Species	Variable sites														
	3	4	9	11	12	15	20	22	28	33	35	36	44	46	47
Man and chimp	V	E	I	I	M	S	V	K	T	H	L	F	P	Y	S
Rhesus monkey	V	E	I	I	M	S	V	K	T	H	L	F	P	Y	S
Horse	V	E	I	V	Q	A	V	K	T	H	L	F	P	F	T
Donkey	V	E	I	V	Q	A	V	K	T	H	L	F	P	F	S
Cow, pig, and sheep	V	E	I	V	Q	A	V	K	T	H	L	F	P	F	S
Dog	V	E	I	V	Q	A	V	K	T	H	L	F	P	F	S
Rabbit	V	E	I	V	Q	A	V	K	T	H	L	F	V	F	S
Californian grey whale	V	E	I	V	Q	A	V	K	T	H	L	F	V	F	S
Great grey kangaroo	V	E	I	V	Q	A	V	K	T	N	I	F	P	F	T
Chicken and turkey	I	E	I	V	Q	S	V	K	T	H	L	F	E	F	S
Pigeon	I	E	I	V	Q	S	V	K	T	H	L	F	E	F	S
Peking duck	V	E	I	V	Q	S	V	K	T	H	L	F	E	F	S
Snapping turtle	V	E	I	V	Q	A	V	K	T	N	L	I	E	F	S
Rattlesnake	V	E	I	T	M	S	V	K	T	H	L	F	V	Y	S
Bullfrog	V	E	I	V	Q	A	C	K	V	Y	L	I	A	F	S
Tuna	V	A	T	V	Q	A	V	N	V	W	L	F	E	Y	S
Dogfish	V	E	V	V	Q	A	V	N	T	S	L	F	Q	F	S

Figure 4.17 Cytochrome *c* sequences of 21 vertebrate species. The whole cytochrome *c* molecule is 104 amino acids long in these species: only the variable sites are shown here; all the other sites do not vary among these 21 species. The amino acids are symbolized by the single-letter code, according to which A = alanine, B = asparagine or aspartic acid, C = cysteine, D = aspartic acid, E = glumatic acid, F = phenylalanine, G = glycine,

The techniques and justification of cladism

50	54	58	60	61	62	65	66	81	83	86	88	89	92	93	95	100	101	103	104
A	N	I	G	E	D	M	E	I	V	K	K	E	A	D	I	K	A	N	E
A	N	I	G	E	D	M	E	I	V	K	K	E	A	D	I	K	A	N	E
D	N	T	K	E	E	M	E	I	A	K	K	T	E	D	I	K	A	N	E
D	N	T	K	E	E	M	E	I	A	K	K	T	E	D	I	K	A	N	E
D	N	T	G	E	E	M	E	I	A	K	K	G	E	D	I	K	A	N	E
D	N	T	G	E	E	M	E	I	A	K	T	G	A	D	I	K	A	K	E
D	N	T	G	E	D	M	E	I	A	K	K	D	A	D	I	K	A	N	E
D	N	T	G	E	E	M	E	I	A	K	K	G	A	D	I	K	A	N	E
D	N	I	G	E	D	M	E	I	A	K	K	G	A	D	I	K	A	N	E
D	N	T	G	E	D	M	E	I	A	K	K	S	V	D	I	D	A	S	K
D	N	T	G	E	D	M	E	I	A	K	K	A	A	D	I	Q	A	A	K
D	N	T	G	E	D	M	E	I	A	K	K	S	A	D	I	D	A	A	K
E	N	T	G	E	E	M	E	I	A	K	K	A	A	D	I	D	A	S	K
A	N	I	G	D	D	M	E	V	T	S	K	K	T	N	I	E	K	A	A
D	N	T	G	E	D	M	E	I	A	K	K	G	Q	D	I	S	A	S	K
D	S	V	N	N	D	M	E	I	A	K	K	G	Q	D	V	S	A	S	
D	S	T	Q	Q	E	R	I	I	A	K	K	S	Q	D	I	K	T	A	S

H = histidine, I = isoleucine, K = lysine, L = leucine, M = methionine, N = asparagine, P = proline, Q = glutamine, R = arginine, S = serine, T = threonine, V = valine, W = tryptophan, X = methylated lysine, Y = tyrosine, Z = glutamine or glutamic acid. (Modified from Dickerson, R. E., 'The structure and history of an ancient protein', Copyright © (1972) by Scientific American, Inc. All rights reserved.)

write an algorithm to discover the most parsimonious phylogeny; if the number of species and characters is large, the number of possible phylogenies becomes astronomical, and no computer can check them all in a reasonable period of time. But the main difficulty is that the principle of parsimony is theoretically suspect. Evolution may not be maximally parsimonious. Indeed, in a trivial sense, we know it is not: 'probably the most parsimonious outcome of evolution would be for it not to occur at all' (Felsenstein 1983a, p. 321). But even given that evolution has taken place, it may still not in fact have been as parsimonious as it could. We have no strong principle telling us to use the method of maximum parsimony; we only have a weak one. It is that evolution is relatively improbable. It is unlikely enough that all the mutations should arise and be selected for one character in one species, but that similar events should take place independently in another is even more improbable. Shared characters are therefore more likely to be due to common ancestry than to convergence. The eye shared by humans and chimps is less likely to be convergent than described from a common ancestor. If evolutionary change is relatively improbable, the method of parsimony will, as more characters are studied, produce the right answer; on this reasonable assumption, the method has the desirable statistical property that it produces the phylogenetic tree of maximum likelihood (Felsenstein 1983b). (Conversely, of course, if more shared characters are convergent than ancestral, the principle gives the wrong answer; on which see Felsenstein 1978.) The principle does not deny that convergence takes place; it cannot be refuted by examples of convergence. Indeed, it is designed to deal with convergence: for if convergence did not happen we should have no need of the principle. The relative improbability of evolutionary change is the valid justification of the principle of parsimony. I should mention that the subject of parsimony is controversial (as the reader of, for example, Gaffney 1979, Cartmill 1981, Panchen 1982, and Farris 1983, can see), and that some invalid justifications have been proposed: that it produces the least refuted theory, that it is needed if we are to do anything at all in the face of conflicting evidence, that it is justified by Occam's razor. Felsenstein (1982, 1983a,b) has dealt with these other proposed justifications, and I may only quote one of his remarks (1983b, p. 271; see also Edwards 1983). 'I am less disturbed by their use [i.e., cladistic methods relying on parsimony] than by the frequent assertion that they make no assumptions about the process of evolution and can claim descent from William of Ockham.'

To seek the tree that requires the fewest evolutionary events is not the only statistical technique justified by the broad principle of evolutionary parsimony. Le Quesne's (1969, 1982) method of 'compatibility analysis' is another. Instead of minimizing convergence, it seeks the phylogeny in which as many characters as possible agree (are compatible), and then fits the other characters to it. And there are other methods, but they will have to be read up elsewhere (such as in Wiley's text, 1981, pp. 176–8, 192, or in Felsenstein 1982) because we are not going to work through them here. None of them can claim to have removed Hennig's dilemma; and, even if they can be justified better than the 'minimum evolution' tree, they at all events rely on arguments very close to the principle of parsimony.

Even though the principle is convincing in some cases, and even though it can be justified, it is not completely reliable. The Hennigian, phylogenetic cladist does not like using it. We only use it because it is often the best principle we have to reconstruct phylogenies from imperfect evidence. Disagreements among characters are so common that almost all our phylogenetic knowledge would be swept away if parsimony were fanatically rejected: we could scarcely say that a human shares a more recent common ancestor with a chimp than with a fish. But, because parsimony is uncertain in principle (and perhaps because it has been justified occasionally by bogus arguments), it is often criticized. The cladist can reasonably retort *tu quoque* to the evolutionary taxonomist (Mayr 1981); but to the phenetic taxonomist (Cain 1967, 1982), cladists should reply that the great interest of phylogenies renders uncertain knowledge preferable to no knowledge, and that the philosophical advantages of cladism relative to phenetic classification outweigh its technical imperfections.

General judgement on the three schools

We can now answer the most important question of taxonomic philosophy: which of the three schools is the best? The answer must be cladism. Both evolutionary and phenetic taxonomists have criticized it, but it can stand up to them. Evolutionary taxonomists (references on p. 32 above) object that cladism ignores a real part of evolution, differential rates of divergence, and that evolutionary classifications are therefore more informative than cladistic ones. (The same claim has been made on behalf of phenetic taxonomy by Cain (1962, p. 11).) The phenetic element displayed in evolutionary classifications is (we are told) real, and important, and part of evolution. A classification that ignores it, according to Mayr 1974 [1976], p. 448), will be 'unbalanced and meaningless'. And 'if we choose to classify purely cladistically (assuming we have enough information) then we fail to display phenetic aspects of nature which are highly important in the evolutionary process, to the organisms themselves, and to ourselves' (Johnson 1970, p. 233). But no matter how important it is, and to whom, its assessment (as Johnson realized) is subjective. For that reason it is unsatisfactory. I therefore do not accept that we must put it in our classifications. Phenetic similarity was rejected from cladism for sound philosophical reasons.

A classification could not possibly represent everything we know about evolution, as evolutionary taxonomists themselves realize. Evolutionary taxonomy itself misses out a real part of evolution, convergence, for its own philosophical reasons. It is not an inherent objection to a school that it misses out part of evolution. Nor is it an advantage to contain many different kinds of information. Indeed, when we start trying to stuff more kinds of information into a classification, we make it less informative, not more. I must therefore exactly contradict Mayr's (1981) defence of evolutionary taxonomy. Mayr accepts that phenetic taxonomy and cladism have made technical advances; but he thinks that they should be confined, in taxonomy, to the place of techniques. The final classification should still be synthetic, still contain both phenetic and cladistic information (for a similar phenetic argument see Sneath and Sokal 1973, and McNeill 1983). Mayr's defence is that such synthetic classifications, because they contain more information, are more informative (likewise Bock 1974). But we must distinguish the amount of information that goes into a classification from the amount that can be taken out. That done, we can turn to Hennig (1966, p. 77, where he cites Bigelow 1956 for the same point; see also Eldredge and Cracraft 1980, p. 201), to discover Mayr's mistake. If species are classified by two sorts of information you do not know, when you are given the classification, which kind of information (or what mixture) was used to define any particular group. You do not know whether any particular relation is phenetic or phylogenetic. The classification may have a high information-content, but it is not informative: it is ambiguous: you cannot extract the information that was put in. However, if the classification is purely cladistic the information is easy to extract. Then the classification is informative. Thus Mayr's argument, if it is properly applied, points away from evolutionary taxonomy. It points, instead, back to cladism.

Phenetic taxonomists originally objected to cladism because they believed that the methods of reconstructing phylogeny revolved a viciously circular argument. Phylogeny could not be reconstructed, and cladism could not represent it. If the characters defining cladistic classifications did not indicate common ancestry, then they must be merely phenetic characters, and cladism would therefore be but a primitive and incomplete form of phenetic classification (Bigelow 1958; Colless 1967; Sneath and Sokal 1973; Sokal 1975). Cladistic classification does use phenetic evidence, and in that sense is a form of phenetic classification; but only in the sense that all classification is phenetic. I should prefer not to call cladism a special case of phenetics. Cladism does not aim to represent a phenetic hierarchy: it aims to represent phylogeny. It chooses characters as indicators of evolution. That is very different from the philosophy of phenetic classification, and is in no sense a special case of it.

Sneath and Sokal (1973; likewise Huxley 1942, allowing for changes in language, Cain 1982, Schindewolf 1968, and many others) do still maintain that the philosophy of cladism is unrealizable, and that it therefore boils down to a special case of phenetic classification; but they have silently softened their criticism from circularity to gross practical uncertainty amounting to ignorance (Cain 1962; Sokal and Sneath 1963; Sneath 1982, p. 209; and see Sokal 1975, p. 259, but not his blind cross-reference to Sneath and Sokal 1973, pp. 109–13 which does not discuss circularity). I only think the criticism has been exaggerated. We do know something about phylogeny. We do not have a perfect knowledge of the phylogeny of all species, but that is not a lethal objection to cladism. Cladism possesses several clearly defined techniques, which can (and have) been applied. Although they are imperfect, they are practical; they are also the subject of active research – cladism is a relatively young school, which holds out the hope of further improvement.

If cladism stands up to its main critics, phenetic and evolutionary taxonomy fall before theirs. They both contain the same fatal error. They have tried to use something in classification, of which there is no natural measure: phenetic similarity. Phenetic classification is inescapably subjective; and evolutionary taxonomy, because it brandishes a partially phenetic philosophy, is also subjective. The phylogenetic philosophy of cladism is objective, and therefore preferable. And the information in a cladistic classification is not only objective and easily recoverable, it is also of particular biological value. The most useful information that a classification can contain, I believe, for a biologist, is phylogeny.

I am not alone as I conclude in favour of cladism. Several other authors, although not all for the same reason as me, also approve of Hennig's phylogenetic systematics: Mayr (1974) cited two authors, Günther (1971) and Griffiths (1972), and I could add Brundin (e.g. 1972), Gaffney (1979), Ghiselin (1984), Wiley (1979, 1981), and perhaps Crowson (1970); I do not doubt there are others. However, it has become difficult to tell how many the numbers are. The word cladism has become ambiguous, as many of Hennig's former followers, while still calling themselves cladists, have abandoned his school to form another of their own: transformed cladism.

5
The reformation of cladism

Cladism, in Hennig's formulation, enjoys an advantage over its two main rivals. It is philosophically more coherent. By discarding the concept of aggregate morphological similarity it has rid itself of the source of subjectivity that plagues, in one form or another, the other schools of taxonomy. In its justification it replaces the phenetic schools' hierarchy of aggregate morphological similarity by the branching hierarchy of phylogeny. The phylogenetic hierarchy does exist in nature, and is unique; cladism is therefore objective in the sense that it seeks to represent a real hierarchy. But to gain that objectivity, it must make the strong assumption that evolution is true. If species did not originate in a branching hierarchy, there would be no phylogenetic hierarchy, and the philosophical coherence of cladism would be lost. Other sources of uncertainty may remain. The recognition of characters and character states is probably an instance, but that is another matter. It is not my argument that cladism is perfectly objective and other taxonomic schools subjective, it is only that the pursuit of the phylogenetic hierarchy, rather than some hierarchy of morphological similarity, makes it objective in one respect. By assuming evolution, cladism gains a philosophical coherence that other schools lack.

This being so, it is strange that, of all schools, it is from within cladism that doubts are being expressed whether evolution is needed in classification. For such is the claim of a new school, whose name betrays its cladistic origins: 'transformed cladism' or 'pattern cladism'. Evolution is unnecessary in phenetic classification, which aims to represent a purely phenetic hierarchy, and claims an advantage over other techniques in the objectivity imposed by its quantitative statistics. This advantage does not depend on evolution. Operational taxonomic units can be clustered without any assumption of evolution. Even the techniques of 'evolutionary taxonomy' might be partially consistent with a non-evolutionary interpretation, for they do after all have a partially phenetic justification. But the cladists!

The reformation of cladism

Cladism, of all schools, must defend its methods by an appeal to phylogeny. And why are they throwing out phylogeny? We have seen that the technical discovery of phylogeny is only uncertain (too uncertain, it might be thought) to support a classification; there might thus be a practical reason for parting with Hennig. But the transformed cladists do not wish to abandon the phylogenetic assumption for practical reasons. They have picked, of all kinds of reason, just the one by which phylogeny puts them at an advantage. They wish to transform cladism for *philosophical* reasons!

It may appear strange; but it has happened. An influential transformed school of cladism has appeared, the doctrine and best-known personnel of which are linked in this passage, written by Patterson (1980 [1982, p. 118]):

> As the theory of cladistics has developed, it has been realized that more and more of the evolutionary framework is inessential, and may be dropped ... Platnick refers to the new theory as 'transformed cladistics' and the transformation is away from dependence on evolutionary theory. Indeed, Gareth Nelson, who is chiefly responsible for the transformation, put it like this in a letter to me this summer: 'In a way, I think we are rediscovering preevolutionary systematics; or if not rediscovering it, fleshing it out.'

To reject the justification of cladism may seem so large a break with Hennig that the school should not be called transformed cladism at all; and Charig (1982, p. 386) has accordingly suggested the term 'natural order systematics' instead. But there are reasons for retaining the name transformed (or pattern) cladism. The new school does retain the techniques of its cladistic origins; and the name has stuck. Transformed cladists themselves do not use the term: significantly they just call themselves cladists. In this work, a transformed cladist means someone who operates the cladistic techniques, but repudiates their Hennigian justification.

The school has more general sounding (and more euphemistic) forms, such as that no theory is necessary in order to discover pattern in nature, or 'pattern recognition should be carried out independently of covering process theories' (Stevens 1983, p. 287, who was only describing, not supporting, the school). And transformed cladism has been magnified by Rieppel (1984) into the latest version of an ancient alternative paradigm to the theory of evolution. He writes (1984, p. 27):

> The modern critique of [evolution][1] is based on the legacy of Cuvier. It consists of the theory of phylogenetic systematics

[1] Rieppel's word was 'transformism', which would be an unfortunate pun here.

(cladism) which is related to the punctuated mode of evolution. The principal aim of cladism is the recognition of the 'natural order' of organisms, much as it was for Cuvier (Eldredge and Cracraft 1980; Nelson and Platnick 1981). The prerequisite of a definite scheme of classification is clearly defined taxa. The resulting order is basically independent of evolutionary theory, but it may be interpreted in an evolutionary context.

He likewise writes (p. 31) that 'cladism is a static view of nature', to be contrasted with the 'dynamic view' of evolution. Just how many cladists are extreme transformed cladists, in the sense of rejecting (or finding 'unnecessary') the theory of evolution from cladism, is not clear. When Charig (published 1982) discussed the school he could cite only two members, and one substantial paper. We have a bit more material now. We shall concentrate on three definite cases: G. Nelson (1978, 1979), C. Patterson (1980, 1982a,b, 1983), and N. I. Platnick (1979, 1982; and Nelson and Platnick 1981, 1984). These three are not the only advocates of the school. Rieppel (1984) could be added to the list; the tone of Rosen, Forey, Gardiner, and Patterson (1981; and Forey 1982; Rosen 1982) suggests another three; Brady (1982) defended the school philosophically; Cracraft (1983, p. 172) mentioned it sympathetically; and, according to Janvier (1984, p. 60), transformed cladism is 'protected by perfect logic'. Stevens (1983, p. 287) also discusses several members of the school. The numbers are difficult to count because transformed cladism is a movement, not a finished school, and different taxonomists accept different parts of the system. But whatever the number of extremists, the influence of the school is not in doubt. Even when it is not directly espoused, its influence can creep into a remark like 'cladistic analysis is essentially the analysis of the relationships among characters (cf. Hull, 1979)' (Meacham 1984, p. 27), in which Hennig's evolutionary philosophy has been silently ignored and the essence of cladism converted into a theoretically unjustified search for natural patterns (I am not implying anything about Meacham's own views). The operator of cladistic techniques might take up any position between the strongly Hennigian philosophy of this book, according to which the theory of evolution is needed to justify the system, and the strongly transformed position. One popular in-between position is to say that the evolution is an 'interpretation' of the cladistic pattern: evolution is then a dispensible, but not dispensed-with, philosophy, which could be taken up and affected, or put down at will; Hennig's strong justification has been ignored. But I do not want here to try to distinguish the degrees of transformation represented by different cladists. Let us concern ourselves no more with personnel. Let us concentrate on ideas. We now know enough members of the school to lead us to the important doctrine.

The main doctrine, as Patterson has just made clear, is that evolution is not necessary to cladism. Before we discuss it, we might notice that some transformed cladists occasionally appear to go further. Nelson (1979) writes about cladograms as if any branching diagram were a cladogram, and 'cladism' as if it were classification itself rather than a school of classification. But he does continue (as do Nelson and Platnick 1981, pp. 140-2, and Patterson 1982a,b) to use a cladistic technique – the embryological criterion – to discover the defining characters of groups; he is still a practical transformed cladist. If, however, any school of classification were ever to stop using cladistic techniques, while still calling itself cladism, it would have moved beyond the transformed cladism I shall criticize here. It would have to be evaluated separately on its merits. In this book, cladistic techniques are those that distinguish ancestral from derived character states (or whatever anyone cares to call them), and a cladogram is a classification, whether or not it is justified as Hennig justified it, whose groups are defined by shared derived characters (or, as they are called, synapomorphies): 'Most cladists [as Funk and Brooks (1981, p. 491n) remark] ... prefer to say that cladistics is classifying based on internested sets of synapomophies.'

If evolution is discarded, the cladistic techniques will have to be reinterpreted. They can no longer be thought to distinguish ancestral from derived character states; for without evolution these concepts are nonsense. The techniques must therefore be supposed to distinguish two kinds of character in some other sense, such that what was formerly called a derived character state is still the kind of character that is needed for classification. (Transformed cladism has for this reason tendentiously renamed ancestral and derived characters 'specific' and 'general'; or retained the Hennigian terms 'apomorphy' and 'plesiomorphy'; or even (Patterson 1982b) raided the term homology for a shared derived character state (cf. above p. 30).) Cladograms will have to be reinterpreted as well: they can no longer express a phylogenetic relation, but must represent only the distribution of cladistic characters. Then 'as summaries of information, cladograms do not in themselves imply any notion of evolution' (Nelson and Platnick 1981, p. 199).

Transformed cladism is undoubtedly operationally practicable. Cladistic techniques can be applied without assuming that evolution has taken place. The reinterpretation can be made. The techniques do not actually, operationally reveal ancestral and derived character states. They were designed to distinguish ancestral from derived character states: the character states that are inferred to be ancestral by a cladistic technique will more often than not indeed be ancestral: but they do not provide the most direct evidence of whether a character

Evolution and Classification

state is ancestral or derived; they rely on indirect inference. Take, for example, the embryological criterion. Suppose that the character state X is judged, by von Baer's law (p. 66) to be ancestral to the character state Y. X is found earlier in ontogeny than Y, and Y develops by the transformation of X. 'In an evolutionary context (wrote Nelson 1978, p. 327), "primitive" means "ancestral." But we would not [by the embryological criterion] have observed character state X to be ancestral in an evolutionary sense. We would have observed only that one character (X) is more general than another (Y), in the sense that X occurs in two species and Y in only one of them.' (Nelson had explained the example in terms of two species.) Platnick (1979, pp. 543-4; Nelson and Platnick 1981, p. 165) makes a similar remark, and a quotation from Patterson (1982b, p. 55; see pp. 50-5 and 1983, pp. 21-2) completes the trinity: 'If "primitive" and "derived" can be dropped, and if our guide is von Baer, not Haeckel (cf. Gould 1977), then belief in, or knowledge of, evolution is clearly unnecessary for the analysis of homologies.'

Such is the transformist interpretation of cladism. Cladistic techniques, it states first, do not actually distinguish ancestral from derived character states. So much is true but not important. Nothing other than a time machine would distinguish ancestral from derived character states on these terms. All that matters for the Hennigian cladist is that the techniques do tend to distinguish ancestral from derived character states. This is not a controversial point. (I can therefore continue to say that the cladistic techniques distinguish 'ancestral' from 'derived' states; the transformed cladist can always translate them.) The theory of evolution, it then says, is not needed to operate the techniques. This is also true, and also unimportant. Any decent set of techniques will be sufficiently formalized that a mindless automaton, which understands no theories at all, can operate them. But the conclusion of the argument, that evolution need not be assumed in cladism, must be opposed.

It ignores, implicitly, a distinction of the profoundest importance. Evolution, in Hennig's scheme, is not needed to apply the techniques of cladism: it is needed to justify them. If a system of classification is to survive in competition with other schools it is not enough to possess a set of working techniques. Almost all taxonomic schools possess a set of working techniques: what we must be told is why one set should be worked rather than another. It is not enough to recognize a pattern in nature; it must explain why that pattern should be recognized rather than any other. It is not enough to believe that, because a system works, justification is not necessary. Platnick (1982, p. 283) attributes to his critics the belief that 'without a causal theory to explain a pattern, the groups do not exist, or (at the very least) are undiscoverable', and replies that groups can and have been discovered without any theory to explain them. But that is not the criticism here: of course transformed cladism can work without the assumption of evolution. The issue here is how it is going to justify itself.

Our search for an answer will fill the rest of this chapter, and all of Chapters 6 and 7. We shall consider four possibilities. The first is that transformed cladism has no justification, and is nakedly subjective. The second is a justification by 'absent' characters. The third, which is the subject of Chapter 6, says that evolution cannot be necessary for classification because classification was performed all right before evolution was ever accepted. The fourth (Ch. 7) says that evolution must be kept out of classification, in order that the facts of classification may be used to test the truth of evolution.

What, then, of the first possible justification, which is to omit one altogether? I have, indeed, found no explicit published justification of transformed cladism. The argument we have been through proves no more than that transformed cladism is possible; and implies that what can be done, will be. It contained no justification. The school has supplied plenty of philosophy, as we shall see, but none of it is in the form of a justification in the sense of this work. One reason for the omission is easily imagined. There is no justification because justification is not possible. The moment the attempt were made the absurdity of transformed cladism would be manifest. Why should we erect a classification defined by (of all things) shared derived characters if evolution is not an assumption? Why pick on shared derived characters? Why go to the trouble of distinguishing ancestral from derived states, and then only using the latter? Why not save the trouble and use another, less eccentric, measure of morphological similarity, such as the various methods of numerical taxonomy? These questions are unanswerable, and completely destructive. In transforming themselves, cladists are reduced to a peculiar faction of phenetic taxonomists, whose idiosyncratic clustering statistic weights each similarly derived characters as 1, and each similarly ancestral character as 0.

Platnick (1979), Nelson (1979), and others (Cartmill 1981; McNeill 1982; Janvier 1984) have realized that transformed cladism closely resembles numerical taxonomy; but not that the whole system is therefore placed in great peril. The only difference Platnick recognizes between numerical taxonomy and transformed cladism is in the interpretation of 'non-congruent' characters. (Non-congruency among characters means that different characters suggest contradictory groupings of species; I have been calling it 'disagreement' or 'contradiction' among characters, see p. 77.) Numerical taxonomists tolerate disagreements among characters, but transformed cladists do not. A numerical taxonomist leaves contradictory characters in the hierarchy, but the cladist always has to go on and show that they are not really proper shared derived characters, that they are based on a mistaken analysis, due to convergence, or to parallelism. But anyone can say that. It makes the difference between a numerical and a cladistic taxonomist little more than a verbal affectation. The cladist says that the minority of noncongruent characters are not really the same characters at all; being convergent, they are really different characters and must be excluded from the classification: the numerical taxonomist leaves these characters in, although their phylogenetic interpretation would be very similar – the pheneticist might sound more sceptical about exactly which characters were convergent, but would not disagree about the fundamental reason why characters fall into contradictory groups.

The real danger of phenetic classification is familiar to us from Chapter 3. It is subjective. If cladism transforms into phenetic classification it will make itself subjective too. It will lose its justification, and become quite arbitrary. And despite the arguments we are about to examine (and they were only implicitly intended to justify the school), transformed cladism does admit that its position is arbitrary and unjustifiable. Thus Platnick (1979, p. 538), at the outset of his paper, gives as two of the three main principles of cladism that 'nature is ordered in a single specifiable pattern which can be represented by a branching diagram or hierarchical classification; second, that the pattern can be recognized by character sampling and finding replicated, internested sets of [shared derived characters]'. He just assumes that his system is true. But anyone can do that. Any numerical taxonomist, with any clustering statistic, can assume that his hierarchy is the true one. Then at the end, Platnick (p. 544, like Patterson 1980 [1982, p. 116]) comes back to his assumption, and tries to justify it not (of course) by evolution, but by metaphysics.

> The refusal to accept noncongruence (i.e. randomness) as a feature of the real world leads us back to what was suggested was the first principle of cladistics: that nature is ordered in a single specifiable pattern. Admittedly, that is not a scientific theory; it can't be tested ... But Popper has argued that such a metaphysical statement, when translated into a methodological rule, is a necessary underpinning of all science.

If Platnick is writing here about the general search for hierarchical classification, then his statement is agreeable. Perhaps we do need metaphysics to justify the search for order. But it is one thing to assume that nature is ordered, and quite another to suppose that it should be classified by shared derived characters. Platnick's second remark may justify taxonomy, but it does not justify transformed cladism. As we compare the second remark with the first we must be quite sure that the justification, in the second remark, for the first principle is not allowed to slip through pretending to be a justification for the second principle. You may be able to call in Popper to justify the search for ordered classifications, but you cannot use him to decide between one clustering statistic and another. The only justification for the second cladistic principle of Platnick is provided not by Popper, but by Hennig.

When transformed cladism throws out evolution, in the absence of any other justification, it immediately makes it choice of cluster statistic subjective. It becomes just one self-consistent system among an infinity of self-consistent systems, with nothing to say in its own favour. If it is consistent at all, it is a naked pact. Effectively the same conclusion has been drawn by previous critics. Charig (1982, p. 372), for instance, concluded that transformed cladism is 'without any proper theoretical basis'; and Ghiselin (1984, p. 109) that the inevitable result of the system is 'some kind of idealism or phenomenalism, in which our personal beliefs about what is homologous or monophyletic become defining of the term' (see also Beatty 1982). A transformed cladist, however, would probably not accept this conclusion, for they do still believe that hierarchical classifications using shared derived characters are superior to other kinds of classification. They have offered three positive arguments in their own favour. Let us now see whether any of them can justify the school.

The first argument would appear to show that it is only possible to classify groups cladistically, other techniques being logically impossible. We might doubt the conclusion from the outset, for we have seen some other techniques in action; but we must examine the argument directly. It is to be found in the remarks of Nelson, Patterson, and Platnick on what they call 'negative' characters, or 'absences'. The section will be quite complex, for we must work through two meanings of the term 'absent' and two arguments for each; of the total of four arguments, two (one for each meaning of 'absent') are evolutionary and valid and two are transformed cladistic and invalid. Let us begin. The obvious meaning of absence is the ordinary language one: 'absence of eyes', for example, would mean that the animal in question lacked eyes. That is not, however, the usual transformed cladistic meaning of the term. There, it usually means not the physical absence of a character, but the absence of a derived character. Fish, for instance, which are a paraphyletic group, are defined by an ancestral character; because characters can only be ancestral or derived, definition by an ancestral character is definition by the absence of a derived character. Thus Platnick (1979, p. 544) on the group: 'if we form a group Pisces, we have based it not on a character, but on the absence of a character. The group Pisces includes those organisms with fins that also happen to lack modified fins (limbs). Such use of the absence of a character is one of the

Evolution and Classification

hallmarks of an artificial group.' Cladists (as we have seen) exclude ancestral characters from their classifications; groups defined by ancestral characters are not allowed. The translation from 'ancestral' to 'absent' therefore can be achieved, with some poetic licence, if we accept the cladistic system, and identify 'character' with 'derived character': then indeed an 'absence' of characters (that is, shared derived characters) do define the kinds of groups not allowed in cladism. The two meanings are completely independent. A character may be absent in the former sense, but not in the latter, and vice versa. Ancestral characters – absences in the cladistic sense – may be physically present (as in the case of fins), or absent (for instance, backbones are absent from invertebrates, which is the ancestral condition before their appearance in vertebrates); vice versa, physical absences may be ancestral (like absent backbones) or derived, as are the lost digits of certain lizards (Lande 1978), or the lost limbs of snakes.

Transformed cladism would not be justified even were it shown that physically absent characters should not be used in classification. Transformed cladistic classifications are defined by shared derived characters, not just physically present ones, and their justification must accordingly show why ancestral characters, not physically absent ones, must be avoided. That point of logic, however, has not stopped transformed cladists from advancing non-evolutionary arguments against the use of physically absent characters; and, for completeness, we must consider them. We shall take the two arguments for the first meaning – physical absence – first. Why should they be excluded from classification? The valid evolutionary reason is that, because information in evolution is easier to destroy than to create, absences are likely to be convergent: it is a case where the functional criterion (Ch. 8) applies. 'Loss convergences' have often been discussed by evolutionary taxonomists. Mayr (1942, pp. 278–9) gives examples from birds, bees, and grasshoppers. Any particular case may be controversial. Birds have often been divided into the flightless 'palaeognathous' ratites and the carinates (most of which fly), but, according to Mayr (1942, p. 278), 'comparative anatomical research has, however, proven, that the Ratites are a polyphyletic group, containing secondarily flightless birds, descendants of probably 5 different flying groups'. The figure of five probably derives from the work of Fürbringer (1888), and although many authorities still favour a polyphyletic theory of ratites (de Beer 1956, 1964), the current of modern research flows in the opposite direction. But even if the ratites are monophyletic (Bock 1963; Cracraft 1974a) and possess ancestral characters relative to the carinates (McGowan 1984), Mayr's general point about loss convergences must stand. Ratites may be monophyletic – although if they are their wings have been reduced on more than one occasion (Cracraft 1974a) – but

The reformation of cladism

flightlessness has nevertheless evolved several times within birds: I need only mention the penguins, the dodo, the takahe *Notornis* of New Zealand, odd ducks, the extinct great auk, and the flightless cormorant of the Galápagos, *Nannopterum harrisi*. Each is probably an independent loss. A phylogenetic taxonomist would not be tempted to define a group by the character 'flightlessness'; the functional criterion would suggest, with overriding power, that the character is liable to convergence.

The functional criterion may suggest that absent characters should not be used in classification, but that justification cannot be allowed in transformed cladism, for it makes double use of the theory of evolution. It both presupposes we are seeking monophyletic groups, and uses the theory of natural selection to argue that absent characters indicate such groups only poorly. Transformed cladism must find another justification. And here it is: 'the character used by von Baer to unite birds and reptiles was "no umbilical cord", that is, no placenta. Aristotle called such negative characters "privative", and argued that they could not logically characterize groups, since they cannot be further subdivided: absence is merely absence (Nelson and Platnick 1981, p. 71)' (Patterson 1983, p. 5; see also Platnick 1982 and Figure 6.1 below for von Baer's classification). However, although absent *characters* cannot be divided, groups defined by them can. The indivisability of absent characters would only matter if groups were divided by successive divisions of their defining characters. But this is not logically necessary, and in modern classification it is generally untrue. Groups within groups are not defined by subdivisions of the same character: lactation (mammals) is not a division of dorsal nerve chord (chordates); and likewise a group defined by the absence of a placenta can be divided by whatever characters differentiate its members. 'In any logic thus far devised there is no problem about propositions concerning absent, or secondarily-modified parts' (Ghiselin 1984, pp. 107–8). If the use of successive division is not a matter of logic, it must be a matter of principle: if it is to be performed, reasons must be supplied. Aristotle did offer reasons, which we shall discuss in the following chapter; they follow from his metaphysics. And if Aristotelian metaphysics is the justification of transformed cladism, then let its advocates say so. I doubt whether it could possibly be justified for the biology of 1980, and am certain that transformed cladism has offered no argument for it. But the truth of the argument can be at most of only incidental interest: it is irrelevant really, because it could not justify their system; it could not justify the use of cladistic techniques. Physically absent characters may be ancestral as well as derived: only arguments concerning the second meaning of 'absence' could possibly justify transformed cladism.

The evolutionary objection to using ancestral characters to define groups is familiar to us: they define paraphyletic groups, and thus fall foul of Hennig's philosophy. The group mentioned in the quotation above from Platnick, fish, are defined by an ancestral character (fins) and, being paraphyletic, are not admitted in Hennigian cladism. But what is the transformed cladistic objection? Let us return to Platnick's remark, to see whether it really gives a valid but non-evolutionary reason for not recognizing paraphyletic groups. Platnick criticized the group because 'if we form a group Pisces we have based it not on a character, but on the absence of a character'. Similarly, transformed cladists have said that groups not defined by shared derived characters are 'uncharacterized', because they are not defined by a character (see, for example, Patterson 1980 [1982, pp. 118–9], 1982a, 1982b, pp. 59, 62; Platnick 1979; Nelson and Platnick 1981, 1984). We seem therefore to be forced to operate the cladistic techniques, not because evolution makes them the only objective method of classification, but because they are the only techniques capable of 'characterizing' groups. Patterson (1981) can thus be left wondering what on Earth an evolutionary classification (Ch. 2) could possibly represent. 'The defining characters of groups like Pongidae and Reptilia are non-existent. The groups have no real existence' (Patterson 1982a, p. 38, and see p. 36 for Pisces likewise). But in what sense are Pisces defined by the absence of a character? They do indeed lack limbs; on the other hand, they do possess fins. Platnick's argument implies that limbs are a character, but fins are not; Patterson's that fins do not exist. In the cladistic translation, fins are not a character in the sense that they are not a shared derived character and cannot define a cladistic group: if we accept cladism, fins might be called a non-character.

But we are not accepting transformed cladism: we are challenging it to defend itself. Its task here is to demonstrate that only cladistic groups can be defined. If Platnick's argument is to justify transformed cladism, it must work only against non-cladistic groups, defined by ancestral characters; it must not work against cladistic groups, defined by derived characters. In fact it looks just as good when stood on its head, although it does point in the opposite direction. Thus, 'if we define the group Tetrapoda (non-Pisces above) we do so by the absence, not the presence, of a character; the Tetrapoda are those organisms with limbs that also happen to lack unmodified limbs (fins). Such use of the absence of a character is one of the hallmarks of an artificial group.'[1] Thus redirected, the argument appears to prove that the cladistically sound group

---

[1] Tetrapoda cannot really be defined by limbs, for snakes, and some other members of the same monophyletic group, lack them. But the absence in snakes is a secondary loss – they are not 'unmodified limbs (fins)' – and pedants may rephrase the definition without altering the point, to something like 'The Tetrapoda are those organisms with limbs or modified limbs ... (where modification can mean loss)'.

Tetrapoda are defined by an absence. The argument fails because the cladistic distinction of ancestral and derived characters is relative. It is therefore possible, in practice, to define groups by presences or absences. In general, if we have (say) five species, three with the derived character state c', and two with the ancestral state c, we can define the group of three either by the absence of c or by the presence of c'. Whether we use presences or absences is completely indifferent. 'How can the absence of something [i.e. an ancestral character] define a group?', Patterson (1980 [1982, p. 114]) asked rhetorically. The answer is, by definition. The Pisces, as Platnick defined them above, are a group of animals. Real animals, which exist in nature, satisfy the definition. Transformed cladists can prove that they would not recognize them as a group; but they cannot prove that a taxonomist who does choose to recognize them is any less wise than themselves. A Hennigian cladist, of course, can, because the Pisces are not monophyletic. But because 'monophyletic' has no non-evolutionary meaning, that argument is not possible in transformed cladism.

In summary, transformed cladism must show that the character choice performed by the cladistic techniques, which is to reject groups defined by ancestral characters, can be justified without any help from the theory of evolution. The two attempts we have met with so far both fail. One sought to show that physically absent characters cannot be used to define groups, but the argument was wrong and in any case inappropriate, for absent characters may be either ancestral or derived. The second argument was at least appropriate, but was invalid. Ancestral characters can in practice be used to define groups. In evolutionary cladism, they should not be; but in that school, the rejection is principled. The distinction of presences from absences does have a certain meaning within cladism, but it is easily confused with the ordinary language meaning, and therefore best avoided; it has none at all in transformed cladism. The transformed cladist cannot invoke it in self-justification.

The transformed application of cladistic techniques has yet to be justified. The difficulty of justification is hardly surprising. The techniques were designed to recognize phylogenetic groups. If you throw out phylogeny, you should throw out the techniques designed to recognize it. But before we draw a final conclusion, let us examine the other two possible justifications of transformed cladism.

6
Classification before Darwin

However species were classified into groups by biologists before Darwin, it was not by the recency of their common ancestry: classification before Darwin was non-evolutionary. Whatever was possible before Darwin must still be possible now, and the advocates of transformed cladism have seized upon pre-Darwinian classification to defend their own school. Thus Nelson, in a remark we have already noticed, wrote: 'I think we are rediscovering preevolutionary systematics; or, if not rediscovering it, fleshing it out' (quoted in Patterson 1980 [1982, p. 118]). Nelson and Platnick (1981, pp. 63–167) and Patterson (1983) have developed the theme. Thus, Patterson remarks (p. 2): 'Today [phylogenetic] trees are usually presented in Haeckel's sense, as pedigrees or images of phylogeny, rather than in the sense of von Baer and pre-Darwinian systematists, as images of hierarchical organization, though cladist systematists have learned to distinguish phylogenetic trees from cladograms.' He then illustrates three models of classification and writes (p. 5), 'the last two examples, and other pre-Darwinian examples ... reproduced by Nelson and Platnick (1981) show that these models have no *necessary* phylogenetic implication'. And then (p. 19), 'the equation of monophyletic groups with natural groups ([three refs]) is another sign of convergence between modern systematics and the pre-Darwinian systematists' attempt to map or mirror a discoverable natural hierarchy' (*natural* here probably does not have the normal meaning of the present book). Similar remarks were made in almost all the works listed on page 88 above.

That Linnaeus, Cuvier, Owen, MacLeay, von Baer, and Agassiz managed to classify without the theory of evolution certainly does show that it can be done. But the practical possibility is not in doubt. What is in doubt is whether cladistic classification can be justified without the theory of evolution. Practice alone is not justification: to justify transformed cladism it is not enough to show that classifications existed before Darwin; it must also be shown that those classifications were cladistic, because otherwise they could only justify non-evolutionary classification in general, not transformed cladism in particular, and that their practitioners had justified their techniques by a valid argument. The methods of Linnaeus and the other pre-Darwinian taxonomists must have been cladistic, their philosophy transformed cladistic. Their justification, moreover, must be valid not only for the biology of 1800, but also for 1980, because we are interested in the validity of transformed cladism now, not two centuries ago. This chapter, therefore, will not be strictly historical. We shall, to be sure, try to understand what earlier taxonomists thought, but only in order to carry their ideas forward into the 1980s, to see whether they are viable now; a sympathetic understanding of the thoughts of earlier ages is not the main aim. To borrow Ghiselin's (1976) distinction, this chapter is not history, but criticism.

It would be disproportionate to cover all the important pre-Darwinian classifications, and we do not need to. We only need to study a sufficient range of classifications to establish ideas that will stand for them all. We shall consider the system of Aristotle, the essentialist classifications of Linnaeus and Cuvier, and the ideal morphology of the first half of the nineteenth century. We shall ask, for each of them, what characters they used in their classifications, why they had chosen them, and whether they classified according to a unique natural hierarchy, or just one of many possible ones. Then we can discover whether they practised a coherent transformed cladism or some revealingly incoherent non-evolutionary method.

Aristotelian classification

In an essentialist classification, species are grouped according to their shared essences, rather than any of their other characters (which are called accidental). Essentialism, therefore, implies character choice – choice of essential characters – but if it is to be practical, it must spell out what the essence of a species is, and how it may be recognized. We can distinguish two kinds of essentialism: the teleological essentialism of Aristotle and the theological essentialism of his later exponents. In the former, the essence of a species X is its *substance,* the *what-it-is-to-be* an *X*. The substantial characters of a species contrast with its accidental ones: without its accidental characters it would still be an X; but without its substantial ones it would not. In theological essentialism, a species substance is the respect in which it resembles (is analogous to) God. The two concepts are related. The theological concept descended from the teleological one: it was developed in those commentaries upon Aristotle, such as by Avicenna and Aquinas, in which the old pagan was adapted to the theologies of the prophet and the Roman Church. And it is easy to see how the development took place. One need only identify the purpose in life of a species with the reason why God invented it, then recollect that God creates species as emanations from himself, and the two concepts collapse into one.

But despite the similarity of the two meanings, we shall be concerned with only one of them here, the teleological version. It is more important to us. It exerted a stronger influence on taxonomists, as we shall see in the case of Linnaeus and Cuvier. They might perhaps have summoned the theological argument to justify the teleological version they practised, but direct thought on the nature of God did not enter into their choice of characters. Moreover, one of the proposed (but inappropriate) justifications of transformed cladism, as we have seen (p. 95),[1] appealed to the authority of Aristotle. Teleological essentialism therefore has a double interest lacking in the theological form. As it happens, we shall cover another form of the theological justification later under the heading of idealism.

We have met the difficulties of teleological classification twice already (pp. 6 and 33); but when we turn to the system of Aristotle we meet another kind of difficulty, a historical one. Historians do not agree, at least in the amount of detail we require, on what his system was, and how (if at all) he operated it. I shall mainly follow Balme (1961 [1975]), who discusses several conflicting authorities (and see Pellegrin 1982). If Balme (1961) is correct, Aristotle did discuss the principles of classification and the characters of animals that might satisfy his principles, but he never formally put his system to work (or no attempt has survived). We therefore cannot discuss his system practically, and are forced into abstraction.

Balme (1961) identified three main rules of classification in Aristotle. The first is the one we have met; that groups must be defined by their substances rather than their accidental characters, where the substance of a species is its purpose in life, the reason why it is what it is. I have said enough about the difficulties of this distinction already. It is difficult, but not meaningless. The characters that define groups are called *differentia,* and in *Historia animalium* Aristotle discusses possible *differentia* in four main classes: bodily parts, life history, activities, and psychological characters. Activities have priority over bodily parts because, in his teleological system, parts are adapted to activities and not vice versa (*P. A.* I, 645b, 14).[1] Once the substance of a group has been identified,

---

[1] *P. A.* = *De partibus animalium.* I quote from the translation of Balme (1972). I have used Ross's edition of the *Metaphysics: The Works of Aristotle,* vol. VIII *Metaphysics,* translated by W. D. Ross (2nd edn, Clarendon Press, Oxford, 1928).

Classification before Darwin

the next step is to convert them into a classification. This is the business of Balme's second rule, according to which the successively smaller subgroups are defined by the successive division of the substance of the larger group. The division must be achieved by dividing the characters that define the higher groups. He makes this clearest in the *Metaphysics* (Z, 1038a, 8-15)

> But it is also necessary that the division be by the differentia *of the differentia*; e.g. 'endowed with feet' is a differentia of 'animal'; again the differentia of 'animal endowed with feet' must be of it *qua* endowed with feet. Therefore we must not say, if we are to speak rightly, that of that which is endowed with feet one part has feathers and one is featherless (if we do this we do it through incapacity); we must divide it only into cloven-footed and not-cloven; for these are differentiae in the foot; cloven-footedness is a form of footedness.

And why must we? Because, if we do not, we may differentiate accidentally rather than substantially. Here is how Balme explains the matter (1961 [1975, p. 184]): '[Aristotle's] reason is that you would thereby be dividing not the genus (footed animals) but another one (feathered animals ...); consequently you could not show that the feathers ... belong essentially to the species in question, because you have not declared these attributes to belong to the genus in the first place.' If being feathered is not substantial in the genus, it cannot be substantial in a part of it.

Another advantage of successive division is its convenience in classification. Without it, a species can possess the differentia of more than one higher group. For instance, if we had divided, at a high level, into *feathered* and *featherless,* and then divided *feathered* non-successively into *tame* and *wild,* a member of the group 'featherless' might well also be wild (or tame). This is not possible with successive division: if we divided the differentia *feather* into different conditions of feathers, no member of the group 'featherless' can possess the differentia. If non-successive division is used, a species can be defined only by listing all its differentia 'in the way that one unifies speech by a connective' (*P. A.* I, 643b, 18), saying 'feathered and tame'; 'tame' by itself does not define the species. Without successive division, as many differentiae are needed as there are divisions; but with successive division, only one differentia, the last one, is needed, because it implies all the others. Aristotle clearly regards this as an advantage of classification by successive division (*Metaphysics* Z, 1038a, 17-24): 'the *last* differentia will be the substance of the thing and its definition, since it is not right to state the same things more than once in our definitions; for it is superfluous. And this does happen; for when we say "animal endowed with feet and two-footed" we have said nothing more than "animal with feet, having two feet."'

Successive division has its own interest, but it also matters to us because it leads Aristotle to his criticism of definition by absent characters (which he calls privations), which we have seen being used by modern transformed cladists in self-justification. The relevant part of Aristotle appears in his critique of what he calls dichotomous classification (*P. A.*, I, ch. 2,3). A dichotomous classification is a simple division into two by a presence and an absence (e.g. *footed* and *footless*). Aristotle disapproves of privations in dichotomous classifications because they cannot be divided: 'there is no differentia of a privation *qua* privation; for there cannot be species of what there is not, for example of *footlessness*' (*P. A.* I, 642b, 21-3). Actually, this criticism is in itself excessive, because, as Balme (1972, p. 109) comments, 'the argument is severe, if not captious ... it is hard to see why Aristotle does not allow us to bypass the privative differentia ... and to continue dividing the higher genus – the animals that remain after footed animals have been marked off. For footlessness in an animal implies other locomotive arrangements.' But however that may be, Aristotle only objects to privations in simple dichotomous classifications, which he objects to anyway: he does not object to them as such. This will become clear after we have seen Balme's third Aristotelian rule.

'His third rule is to divide the genus by a plurality of differentiae, not by one at a time' (Balme 1961 [1975, p. 184]). No single differentia, even if it is a positive character, will have enough divisions to distinguish all the species. 'This is why one should divide off the one kind straight away by many differentiae' (*P. A.*, I, 643b, 23). But if many differentiae are used the objection to privations disappears, for the group of differentiae (including privations) can be successively divided: 'In this way too the privations will make a differentiation, whereas in dichotomy they will not' (*P. A.*, I, 643b, 25). In the system Aristotle recommends, he admits privations; only in the system he rejects does he object to them.

So much for Aristotle. Despite the appeal of Nelson, Platnick, and Patterson, he did not justify transformed cladism. Privations, for a start, are not ancestral characters, and even if they were the Aristotelian argument would not help. The criticism of privations applies to classifications with single differentia; but Aristotle recommends the use of multiple differentiae. It also presupposes successive division. Transformed cladism, therefore, must find a justification of successive division, and of division by single differentia, before it can appeal to Aristotle. There may be a justification of successive division in Aristotle's metaphysical distinction of substance and accident; but some work would have to be done, for that distinction is not obviously acceptable in modern biology. As for a justification of simple dichotomies of single differentia, I can only suggest that one might be discoverable in the Platonic classifications Aristotle was criticizing; but I shall leave it to others to explore that mysterious territory.

The transformed cladistic appeal to Aristotle, therefore, does not work. It is not enough to confuse ancestral characters with privations, and then fetch an inappropriate justification, from a system that is at best obscure and is in fact probably no longer tenable.

Aristotelian essentialism in Linnaeus and Cuvier

Linnaeus, if we concentrate on his botany, classified with the characters of the reproductive system. Why did he pick them? Because they were essential characters, of high importance in the life of the plant. They also met the requirement of Aristotle's third rule, for they allow many differentiae. (Sloan 1972 credits this discovery concerning reproductive systems to the botanist Cesalpino (1519-1603).) Linnaeus's classifications were thus essentialist in the Aristotelian sense (Cain 1958). The same kind of character choice can be seen in the classifications of Baron Cuvier, which, if I understand them correctly, also belong in part in the Aristotelian tradition. Cuvier summarized his philosophy of classification in the introductory essay to *La règne animal* (1829 edn).[1] He clearly distinguishes technique and justification. The natural classification of animals into classes, orders, families, and so on, he, like his contemporaries, calls, *une méthode* (p. 8). He discusses what the best method would be. It would be essentialist:

> When the method is good, it does more than teach names. If the subdivisions have not been established arbitrarily, but are based on true and fundamental relations, on the essential similarities [*rassemblances essentielles*] of beings, the method is the surest means of reducing the properties of beings to general rules, of expressing them in the fewest words, and of fixing them in the memory.

[1] Which I have only ventured into with the help of Cain's (1959a) commentary. According to Cain (1959a, p. 186), the classificatory philosophy of de Candolle was similar to Cuvier's. I should add that I now interpret Cuvier very differently from Cain, whose own empirical philosophy has led him, I believe, to underrate the importance of Cuvier's theory. The quotations from Cuvier come from Cuvier (1829, pp. 8-10), which work was translated anonymously in 1834; I have slightly altered the translations.

Evolution and Classification

The classification described by Cuvier is natural in the usual sense of this book: the groups share other properties besides their defining character: the classification is 'the best means of covering the properties of beings by general laws'. Cuvier does not strictly offer this in justification of his method, although he clearly thinks it a desirable property. But in any case, we have already seen that simple phenetic 'naturalness' cannot support an objective classification. Cuvier says that a 'good method' gives natural classification, but what is the 'good method' of recognizing characters? Cuvier explained all correlations of characters by functional necessity. The characters that would be correlated with the greater number of other characters would be the functionally most important ones, which, accordingly, should be used to define groups.

> We assiduously compare beings, under the direction of the principle of the *subordination of characters,* which is itself derived from that of the conditions of existence. The parts of a being must all be mutually adapted; some traits exclude others, others necessitate them: when, therefore, we recognize one such trait in a being, we can calculate in advance those that will coexist with it, and those that will not. The parts, properties, or traits that conform with the greatest number of others, or, in other words, that have the most marked influence on the whole of the being, are called the *important characters,* the *dominant characters*; the others are called the *subordinate characters*; and there are degrees of each.

And how are the dominant, important characters to be discovered? Cuvier gives two methods:

> This influence of the characters is sometimes determined rationally, by considering the nature of the organ. When this is impossible, we use simple observation, and a sure means of recognizing important characters, one derived from their own nature, is that they are the most constant, and that in a long series of diverse beings, arranged according to their similarity, these characters are the last to vary.

Cuvier, therefore, like Linnaeus, sought the essential characters of organisms to classify them. He wished his classifications to be natural in the sense that they would cover many characters; he believed that functional integration was the reason of correlations among characters; he therefore chose characters of the greatest functional importance. This method looks like the exact opposite of evolutionary taxonomy, but in fact the two schools are similar. They appear to differ because, although both aim to identify functionally

Classification before Darwin

important characters, Cuvier appears to do so in order to define groups by them, and evolutionary taxonomy in order to avoid doing so. The difference, however, is more one of interpretation than practice. Both schools classify with characters that show little variation; they therefore define similar groups. Cuvier did explain the invariance of characters by their functional necessity, in contrast with the evolutionary taxonomists's talk of broad adaptation, or even non-adaptation; but Cuvier's philosophy could probably be incorporated into evolutionary taxonomy. It would be easy if functionally important characters were likely to be broad adaptations. The philosophy of evolutionary taxonomy could also be imported into Cuvier's method: Cuvier sought characters that would be likely to indicate natural groups, which is just what the characters of evolutionary taxonomy are meant to do.

Cuvier's technique of character choice can be seen at work in his classification of vertebrates (Cuvier 1829, on which see Cain 1959a, pp. 200–1). What are the important characters here? He recognized four groups of vertebrates, 'characterized by the kind or power of their motions, which themselves depend on the quantity of their respiration'. The differences in the method of locomotion was the most important character, and the circulatory system closely associated with it. They were therefore used to divide the vertebrates (Cuvier 1829, pp. 56–9; Cain 1959a, pp. 200–1, especially 201n for Cuvier's meaning, now exceptional, of 'single' and 'double' circulation):

1. Mammals. Double circulation. Aerial respiration in lung only. Strong walkers and runners.

2. Birds. Double circulation. Aerial respiration in lung and air sacs. Fly.

3. Reptiles. Single circulation. Aerial respiration in lung. Crawl.

4. Fish. Double circulation. Aquatic respiration in gills. Swim.

He put the same principle to work on each group. For instance, 'the variable characters that establish the essential differences among the mammals' (Cuvier 1829, p. 65) are the organs of touch, because they determine dexterity and skill; and 'the organs of *manducation,* which determine the nature of the animal's food, and entail not only everything relating to digestion, but also a mass of other differences, even including intelligence' (p. 66). These two classes of characters divide the mammals into nine orders (Cuvier 1829, pp. 67–9). The first five have separate fingers on their limbs. Of these five, the first three have three sorts of teeth, grinders, canines, and incisors: Bimana (i.e. man, hands on two limbs); Quadrumana (hands on four limbs); Carnaria (non-opposable thumb). The second two are the Rodentia (which lack canines), and then 'those animals whose toes are cramped, and deeply sunk in large nails, which are generally curved; they have no incisors ...', that is, the Edentata. Then there are two hoofed orders, Ruminantia and Pachydermata; and finally, Cetacea, which lack posterior extremities, and Marsupialia. The classification illustrates, by the way, that Cuvier was not a full Aristotelian: the successive subgroups are not defined by successive division. Vertebrates were defined by the circulatory system, mammals by their feeding mechanisms.

Cuvier and Linnaeus classified by characters of functional importance. They were essentialists in the teleological, Aristotelian sense. As such, their philosophy was neither phylogenetic nor phenetic; but whether such a philosophy could be formulated validly for the 1980s is (as we have seen) an open question. At all events, it could not justify the non-evolutionary application of cladistic techniques, for it does not lead to the cladistic technique of character choice. The characters chosen by the teleological essentialist differ from all three modern schools. In interpretation, if not practice, essentialist classification is the exact opposite of an evolutionary classification. It differs from cladism, because, except by coincidence, functionally important characters need not be shared derived characters. Aristotelian classification differs from numerical phenetic classification because Aristotelian taxonomists choose their characters: they exclude characters judged accidental or subordinate or unimportant; the numerical taxonomist, by contrast, does not choose particular kinds of character at all. The technique of teleological essentialism is not cladistic, its philosophy is not transformed cladistic. The fact that it was practical, in pre-Darwinian times, should provide neither comfort nor justification for that school.

Morphological idealism: secular and divine

Cuvier's morphology, according to which each part of an organism is well designed for life, would have been more familiar in Great Britain in the 'argument from design' of the school of thought for which Paley's name may stand. Its influence did persist through the century, but by 1850 it had already become old-fashioned. The years from Cuvier to Darwin, and beyond, saw its replacement by what may be called idealist morphology (Bowler 1977, 1983; Hull 1973; Ospovat 1981; Yeo 1979; Ridley 1982; Rehbock 1983; Russell 1916 is the classic account of nineteenth-century morphology). Idealism, in this context, is the doctrine that nature is constructed from ideal forms; in morphology, it becomes the doctrine that the forms of animals are variants of a limited number of plans (or ideas). Species

Classification before Darwin

will then show similarities because they embody the same idea, rather than because they have a similar functional design. Morphologists thus came to emphasize the characters of organisms that (they believed) were non-functional but appeared to conform to the 'plan' of the group: male animals, for instance, possess useless nipples, which are part of the mammalian plan, and whales possess a set of useless bones planned on a hind limb. Vestigial organs are positively functionless, but the same theory would also point to similarities – such as between the skeletal designs of a horse and a man (Owen 1848, pp. 11-14, on which see Cain 1964, pp. 40-1), which do not appear to be functionally necessary: if a horse and a human had been independently designed they need not have shown such detailed similarity of form.

Transformed cladists, as it happens, have been keener to find their ancestors among idealists such as Agassiz and von Baer than essentialists such as Linnaeus and Cuvier. The present section is therefore perhaps more relevant to an analysis of transformed cladism than was the previous one. As usual, we must proceed by asking whether the philosophy, idealism in this case, can justify an objective school of classification and, if it can, what kind of character choice (and what taxonomic techniques) it implies. If it is to justify transformed cladism, idealism must offer a valid non-evolutionary objective principle of classification by shared derived characters. Let us see if it does. Classification, according to idealist morphology, should represent the plan of nature. Is there any reason to suppose that nature has a unique, ideal plan? And if there is, by what techniques should it be discovered?

In answer to the first question, there were I think three reasons. Two of them are mentioned by Agassiz, in his *Essay on Classification* (1859). The one favoured by Agassiz (1859, p. 8) himself is that 'the divisions of animals according to branch, class, order [etc.] ... have been instituted by the Divine Intelligence as the categories of his thinking'. Perhaps Agassiz's idealism was more theological than that of other morphologists: but MacLeay (1825, p. 47) said he would classify by 'the plan by which the Deity regulated the creation', Owen (1848, p. 73) by the plan 'on which it has pleased the divine Architect to build up certain of his diversified living works', and, according to the historian Ospovat (1981, p. 102), 'the supposition that the natural system was God's plan was widely shared'. It was perhaps less often expressed in the lands of *Naturphilosophie* or *anatomie transcendente* than of natural theology; they may have preferred the other reason for a natural system of types, which Agassiz cast down into a footnote:

> It must not be overlooked here that a system may be natural, that is, may agree in every respect with the facts in nature, and yet not be considered by its author as the manifestation of

the thoughts of the Creator, but merely as the expression of a fact existing in nature – no matter how – which the human mind may trace and reproduce in a systematic form of its own invention.

In theory, this second philosophy could come in two forms, naively empirical or Kantian; that is, the natural classification might be said simply to exist in nature (naive empiricism) or we might be forced to discover it because our minds are built the way they are (Kantian). The difference, however, need not concern us. We can take the two empirical philosophies together, as versions of phenetic taxonomy. According to both, nature has a single hierarchical plan, defined by phenetic similarity alone.

We know how to deal with that philosophy. A biologist in the early nineteenth century might, quite reasonably, believe in a natural phenetic hierarchy. I do not know how many biologists then held such a belief: my impression is that most believed also that the hierarchy was divinely underwritten; but many did believe the natural plan could be discovered by simple observation. The theory, however, is no longer reasonable. It was finally exploded by the critics of numerical taxonomy. Nature does not have a unique phenetic plan. Agassiz, Owen, von Baer, could discover a plan of nature all right, but it would be one plan among many; the techniques by which they defined groups must have been chosen (from this broad perspective) subjectively. Their classifications can no longer be justified by an 'empirical' philosophy.

If the empirical philosophy must be abandoned, what about the divine one preferred by Agassiz? It differs from what I called theological essentialism. In both systems, the groups of nature are thoughts of God; but in (for instance) Aquinas's analogical essentialism, the groups are also analogies of God: the plan of nature is not only a thought of God, but God is thinking about himself. Aquinas did prove that God can only think about himself (God is the only perfect being; a perfect being can only think perfect thoughts; therefore God can only think about himself), but his argument is theologically contestable (Passmore 1970), as God arguably could think about a natural plan somewhat independently of self-contemplation. The idealism of Agassiz says only that the plan of nature owes its existence to divine thought; it does not have to adopt the essentialist corollary that God is *ipso facto* thinking about himself. But that theological distinction is not important here. All we are concerned with is whether Agassiz's divine philosophy can justify his system of classification. To do so, it (and any modern transformed cladist who seeks comfort in the classifications of Agassiz) must show three things: first, that God exists; second, that at any time God thinks unique thoughts; and third, that God's thoughts are discoverable. The detailed nature of the divine thoughts would then lead to a technique of character choice in classification. If Agassiz is to justify transformed cladism as well, he must also have demonstrated a fourth proposition, that God thinks *cladistic* thoughts. Let me comment on each of the four questions. The first is difficult and uncertain, and too large to enter into here; but the existence of God does sound a strange premise for a system of classification for the 1980s. The second, however, has been settled. The very first Christian philosopher, Philo, a Jew of Alexandria and a contemporary of Christ, proved in his work *On the Unchangeableness of God*, that God can never change his mind; and later thinkers have agreed with him. The third question is the most important for us. Can we discover the divine thoughts, and thus find the proper principle of character choice for biological classification? Here, indeed, is a difficulty. A school of classification, such as idealism, will employ a principle of character choice: but it should also tell us why, if God thought up the classified plan, did he choose *those* particular characters? Neither Agassiz, nor anyone else (so far as I know) has even tried to answer this question. Agassiz undoubtedly thought God caused the plan; but he did not try to show that God caused the plan Agassiz chose to classify. Agassiz merely assumed it, and rushed on to the consequence that pleased him so much, namely, that if God thinks of the plan of nature, then as the taxonomist himself discovers the plan, 'his mental operations ... approximate the workings of the Divine Reason'. The taxonomist becomes divine. He feared 'such a suggestion may appear irreverent', but concluded it was not: he was, after all, writing in Boston, not Rome. For all the argument Agassiz gives to justify his choice of characters, almost any mental operation in a taxonomist could be divine. His character choice, in other words, was subjective. He had not thought about the problem.

Perhaps it was not unreasonable of him not to. He could just assume, reasonably if naively, that the plan of nature he discovered was the divine one. He thought that, as a matter of fact, there was only one phenetic hierarchy, not an infinity of them. The question of whether his plan was the divine one would not have occurred to him. It is an anachronistic question; but nonetheless necessary for our purposes. If Agassiz is to be the prophet of any modern school, he must have answered the questions of modern critics: he must have revealed why God thinks only about particular kinds of characters; otherwise we cannot know whether any particular classification corresponds to the divine plan, and (like Agassiz) we can only use divine idealism to interpret classifications that have been discovered with techniques of secular inspiration. The interpretation is built on the classification; it does not justify it. This being so, the theology is quite superfluous. It has no use in classification. Darwin, indeed,

Evolution and Classification

used to complain about this at the time: 'but many naturalists think that something more is meant by the Natural System; they believe that it reveals the plan of the Creator; but unless it be specified whether order in time or space, or what else is meant by the plan of the Creator, it seems to me that nothing is thus added to our knowledge' (Darwin 1859 [1969, p. 399]: Gillespie 1979 discusses the history). But, as Darwin was probably aware, in natural theology biology was supposed to add to our knowledge of God, not vice versa. Theology was not intended to justify a school of classification; if anything it was the other way round. No wonder it cannot justify transformed cladism.

What should we conclude? From the authority of Philo and his followers, we can say that, if God does think about the plan of nature, he will think unique and unambiguous thoughts. Agassiz's philosophy, therefore, could in theory justify an objective system of classification. But the conclusion is theoretical only. Ideal morphology is in the position that the critics of cladism attribute to that school (see pp. 84–5): it may have an objective principle, but it cannot be realized because it lacks satisfactory techniques. But whereas we do have some techniques to discover evolution, morphological idealism really is practically useless. We have no idea how to discover the thoughts of God. The justification works only if we can show that our techniques do discover the divine plan: without that argument, our techniques are unjustified, and our classification subjective. Because we do lack such an argument, the divine interpretation cannot justify morphological idealism.

Nor can it justify transformed cladism. To do so, it would have to show that God, in his biological faculty, thinks *cladistic* thoughts: that he contemplates shared derived characters only. That has not, and (surely) cannot, be shown. The fate of morphological idealism, in its empirical and divine forms, only proves still further that no phenetic or 'empirical' system of classification can be justified. When it is performed, it must be (from our modern viewpoint) subjective.

So much for the philosophy of the ideal morphologists. What about their techniques? They may not have been objective, but they did exist. If Owen, MacLeay, von Baer, and Agassiz, were ancestral transformed cladists, their techniques must at least have been cladistic. But were they? The quickest route to an answer is to search their classifications for non-cladistic groups. If they were cladists, they should not contain such non-cladistic groups as fish, or reptiles, or primates. The third chapter of Agassiz' *Essay,* 'Notice of the principal systems of zoology', is a synopsis of the classifications of the time. Let us turn first to the 'embryological system' of von Baer. Does it have reptiles and fish? Indeed it does: there they are, on page 360. And what of the 'physiophilosophical system' of MacLeay? That too has Pisces and Reptilia, on page 347. So does Owen, in his

Classification before Darwin

'anatomical system', on page 326. And so on. Agassiz summarizes 18 systems in all, from Linnaeus to Vogt, and they *all* possess the tell-tale non-cladistic groups: not one of them, it appears, was cladistically defined. Not one of them, even though they were all pre-Darwinian, was a transformed cladist.

The classifications produced by von Baer and others, however, are only prima-facie evidence. They are not conclusive. Cladism is a method rather than a classification, and if von Baer (for example) had defined the reptiles by a shared derived character, then although we should now think him mistaken (for Reptilia cannot be so defined), he should be counted as a transformed cladist. He could have the method right, but bungled its application. We must therefore look behind the groups, to see how they were defined. Figure 6.1 reproduces von Baer's divisions of the Vertebrata, and reveals two groups of fish, the Urodela, the Anura, and the Reptilia. They are all defined by negative, absent characters. Absent characters can be derived (p. 95 above), but all of them in this case are ancestral. He even partly defines the birds by an ancestral absence. The same general point is true of the other taxonomists summarized by Agassiz. Their groups were neither cladistic, nor defined cladistically.

If the morphological idealists were not forerunners of cladism, what were they? Of our three main schools, morphological idealism was closest to evolutionary taxonomy. Evolutionary taxonomy (Ch. 2) defines groups by homologies, and seeks to recognize and exclude analogies. Homologies were first distinguished from analogies by morphological idealists, and for a comparable purpose (Swainson 1835; Owen 1848; Agassiz 1859). They too classified by homologies, not analogies. Homology, of course, did not then have its modern evolutionary meaning. But it was still a principle of character choice. A principle must distinguish the true characters of organisms, which may be used to define groups, from the false ones. The homologies were the true characters, because they realized the ideal plan of nature; analogies were not part of the plan. Analogous similarities were imposed by functional demands. Homologous similarities were those not so imposed; they therefore must exist for some other reason: the plan of nature.

Owen's great achievement was to abstract all the homologies of the vertebrates to reveal the vertebrate idea, or 'archetype', in its true form. The archetype was an epitome of all the homologies – all the true morphological characters of vertebrates – which (as we have seen) Owen interpreted theologically. Different idealists did recognize homologies by different techniques. Agassiz preferred embryological characters (1859, p. 129). Owen did not: he criticized embryology, and preferred to compare the anatomy of adult bones (Owen 1848, pp. 104–6). But they shared the same aim, which was to classify according to the ideal plan, which was recognized by homologous similarities.

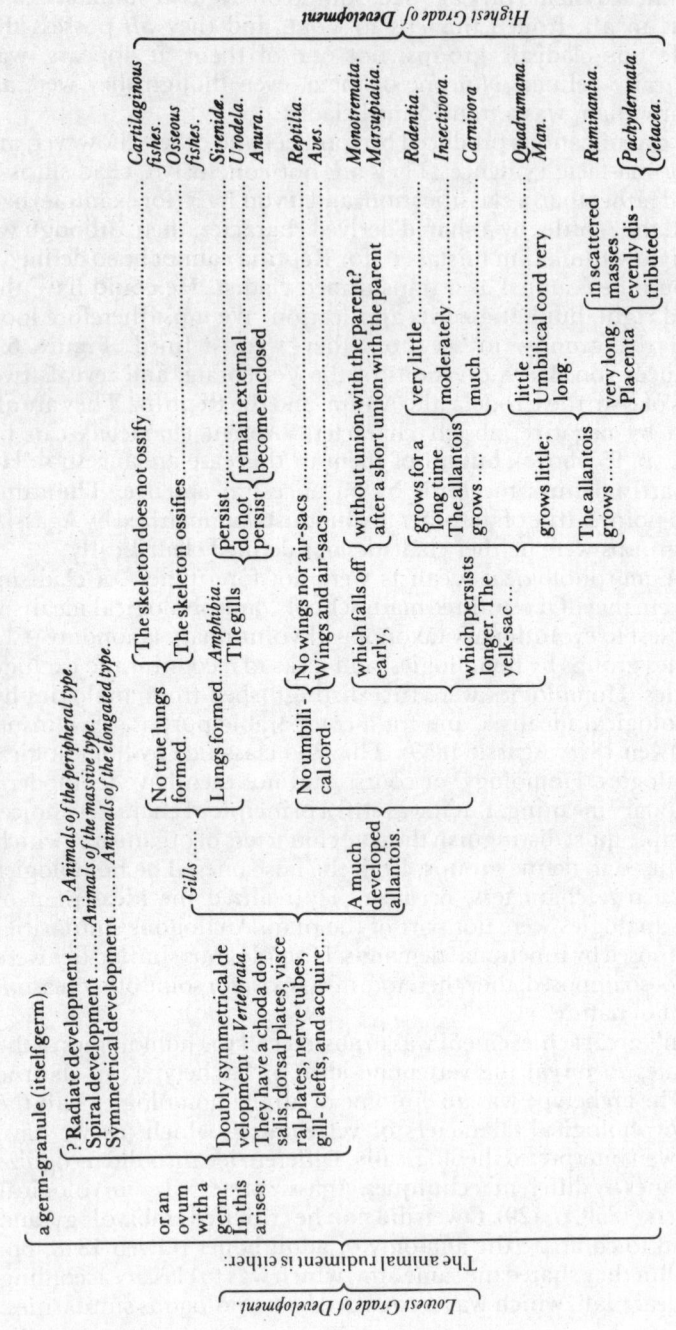

Figure 6.1
Von Baer's divisions of the vertebrates (from von Baer 1828 [1853, p. 215]).

```
Scheme of the Progress of Development.

                                                                                              Highest Grade of Development
                                          ┌ The skeleton does not ossify ........... ┌ Cartilaginous fishes.
                                          │                                          └ Osseous fishes.
                                 ┌ No true lungs  ┤ The skeleton ossifies ............... ┌ Sirenidae.
                                 │ formed.       │                                        │ Urodela.
                                 │               │                                        └ Anura.
                          Gills ─┤
                                 │               ┌ persist ................................ Amphibia.
                                 │ Lungs formed ─┤         ┌ remain external ............. Reptilia.
                                 │               │ do not ─┤
                                 └ The gills     └ persist └ become enclosed ............. Aves.
  
  Doubly symmetrical de-        ┌ No wings nor air-sacs ...................................... ┌ Monotremata.
  velopment ... Vertebrata.     │                         ┌ without union with the parent?     └ Marsupialia.
  They have a choda dor-        │                         │ after a short union with the parent
  salis, dorsal plates, visce-  │ Wings and air-sacs      │                                    ┌ Rodentia.
  ral plates, nerve tubes,      │ which falls off         │ grows for a ─ very little ........ │ Insectivora.
  gill- clefts, and acquire...  │ early,                  │ long time    ─ moderately....      │
                                │                         │ The allantois ─ much .............. └ Carnivora.
                                │                         │ grows.....
                                │
                                │                         ┌ grows little; ┌ little ............. ┌ Quadrumana.
                                │ which persists          │               │ Umbilical cord very   └ Man.
  A much                        │ longer. The             │ The allantois │ long.
  developed                     │ yelk-sac...             │ grows...      │ very long. ┌ in scattered   ┌ Ruminantia.
  allantois.                    │                         │               │ Placenta   │ masses.        │
                                │                         │               │            │ evenly dis-    ┌ Pachydermata.
                                │                         │               │            └ tributed       └ Cetacea.

  a germ-granule (itself germ),   ? Radiate development ...... ? Animals of the peripheral type.
  or an ovum with a germ.         Spiral development ........... Animals of the massive type.
  In this arises:                 Symmetrical development ..... Animals of the elongated type.

                                  Lowest Grade of Development.
                                  The animal rudiment is either:
```

Although homologies were linked, in the idealist system, to an idealist philosophy, it identified much the same characters as homologies as an evolutionary taxonomist would now: Owen, for example, described the wings of birds and bats as analogous (1848, p. 7) and tetrapod limbs as homologous (pp. 5, 7). It was therefore possible for morphologists who lived through, and accepted, the Darwinian revolution, to continue business as usual (Coleman 1976). Evolutionary taxonomy has stripped the homology – analogy distinction from its idealist philosophy; but it still owes the distinction itself to that now discarded school of thought. If any school were to seek comfort in pre-Darwinian idealism, it should be a (non-existent) 'transformed' evolutionary taxonomy, not transformed cladism.

The purpose of this chapter was to discover whether the techniques and justifications of classification before Darwin could justify the system of transformed cladism. It has been successful in its purposes: it can conclude unhesitatingly in the negative. Pre-Darwinian classification does not justify transformed cladism. We have examined two pre-Darwinian systems of classification, essentialism and idealism. In order to justify transformed cladism, they would have to operate cladistic techniques, and provide a valid (in modern terms) justification for doing so. They do not do either. They are both utterly different from transformed cladism, their only similarity to it being a negative one, that they would not justify themselves by the theory of evolution. In all other respects the schools differ. Neither essentialist nor idealist classification uses cladistic techniques. Essentialism, in its teleological form (as practised by Linnaeus and Cuvier), groups species by shared purposes in life and thus defines groups by characters of high functional importance; idealists defined groups by homologies, which were thought to embody a transcendental plan of nature.

Nor can essentialism or idealism justify non-evolutionary classification even with the characters they do use. The essentialist justification is incomplete, and we cannot say whether it could be completed. Even if it could be, it would have to be an evolutionary principle, because natural selection gives organisms the only purposes they have. The status of idealism is easier to settle. Its naive empirical form is nakedly subjective: its divine form is impracticable: it can therefore be rejected in both its forms. The favoured form was the divine one, according to which the ideal plan was supposed to exist because God was thinking it. Because the thoughts of God are undiscoverable, all an idealist could do was to pick characters subjectively, and then interpret the groups, after the fact, as the thoughts of God. As it happens, they chose characters very like our modern homologies. Transformed cladism, if it really were to 'rediscover pre-evolutionary systematics', could only converge on idealism by defining groups subjectively, with shared derived characters, and then declaring that these exist in the mind of God. No one will be able to prove them wrong. We do not know whether God thinks about the Reptilia: Agassiz and his contemporaries thought he did; the modern transformed cladist would maintain he does not. The conflict could never be resolved. The system is subjective. The chapter therefore confirms the existing conclusion. The only known objective, practical system of classification is Hennigian cladism. In it, the cladistic techniques are tied to the theory of evolution. If evolution is rejected, there is no reason to keep the cladistic techniques. Now we have looked at some pre-evolutionary methods, we have seen that they are, or boil down to, techniques that are subjective; we have also seen that they were not cladistic.

But perhaps this chapter has taken the advocates of transformed cladism too seriously. Perhaps the pre-Darwinian transformed cladists are meant only to be a comfortable myth, not a critical justification of the modern school. For most revolutionaries do invent historical myths about their ancestors. The agitators of the Great Rebellion encouraged themselves with the idea that ancient parliamentary liberties had been crushed beneath a 'Norman yoke', from which they would now, at last, liberate themselves (Skinner 1965). After the Calvinist revolutionaries of Scotland had forced their Queen to abdicate, their ideologists showed it was quite all right, for under the 'Ancient Scottish Constitution' the Scots were doing that kind of thing all the time (Trevor-Roper 1966). And now revolutionary cladism would portray itself as the rescuer of a good old cause from the oppressions of a Darwinian yoke. In all these cases the history is equally mythical. The Anglo-Saxon Witan was scarcely a sovereign parliament; the wicked kings of Scotland never lived; and the ancient transformed cladists were neither 'transformed' in philosophy, nor cladistic in their method.

Our conclusion need not only be negative. The historical analysis has extended the argument in two respects. In the essentialist principle, we met again a possible alternative to the phylogenetic and phenetic principles. There may be other objective principles besides the phylogenetic one, and teleology may be one of them. The theological form of morphological idealism illustrates another point. An objective principle of classification must be practical. We may admit that the thoughts of God fall into unambiguous categories. But because we have no method of their discovery, the principle is practically useless. The phylogenetic principle is practical as well as objective. We have techniques by which it can, imperfectly, be discovered. But if they did not discover evolution, they would not be objective; they are tied by logic to the theory of evolution. And as this chapter has now shown, they are not only linked by logic, but by history too.

7
Classification and the tests of evolution

The third proposed justification of transformed cladism is as a test of the theory of evolution. Evolution must be excluded from classification in order that the classifications may then be used to test the truth of evolution. The argument has not been discussed in any detail in the literature of transformed cladism; but it has been mentioned occasionally in print, and is often raised in spoken form. Here it is from the pen of Platnick (1979, p. 539): 'if classifications (that is, our knowledge of patterns) are ever to provide an adequate test of theories of evolutionary processes, their construction must be independent of any particular theory of process'. The same argument can be found in Rieppel (1980) and in Janvier (1984, p. 70); it is approved by Thomson (1982); and it fills the concluding sentences of one of Patterson's essays (1982a). He writes:

> As a science of pattern, cladistics holds out the possibility of a reconstruction of the history of life in space and time that does not depend on Darwinian or neodarwinian presuppositions. The interest of that reconstruction or cladogram is that theories of process – neodarwinian or any other – can be tested only against nature... But if we are taught, as we have been, to see that pattern through the spectacles of evolutionary theory, how could the pattern ever test the theory?

The essay ends there, with the clear implication that no answer can be supplied. These passages do suggest that transformed cladism has been justified as a test of evolution, but they should be used only tentatively: they are short comments, not formal arguments; and they were not actually presented as explicit justifications of transformed cladism, although Patterson does call it '*the* interest' of transformed cladism. The limited number of references hardly show that the argument is crucial to transformed cladism; but then it is difficult to determine just what are its crucial arguments, for, as Charig (1982, p.

Evolution and Classification

385) has also found, its position is 'constantly shifting'. I do not object to conceptual development; but, if transformed cladism is not a static body of doctrine, it does not much matter whether this chapter criticizes an idea believed by particular people. All that matters is that the ideas themselves should be important. And they are: they are not new to the controversies of classification; in one form or other, they have been with evolutionary taxonomy throughout its history (Hull 1967), as we have seen in the case of the numerical taxonomic critique. I do believe that many biologists think that transformed cladism can be justified as a test of evolution: I may be wrong; but even if no one believes it, even if this chapter contains nothing to frighten a transformed cladist, it remains a real discussion of a real question. And then, even if some readers do think this chapter unfair, others will probably agree that the argument should be considered.

The argument itself is clear enough. If asked why we think species evolved, rather than being separately created, we should probably include classification in our reply; any textbook certainly would. A transformed cladist may then reason that, if evolution was an assumption of the classification, surely the classification could not then test evolution: if we wish to test evolution, we must extirpate it from classification; and that (he would conclude) is why cladism must be transformed. It looks as if there may be a justification here. If a classification is built without the help of the theory of evolution, and then turns out to fit the theory, we should have clear evidence for evolution. Since the truth of evolution is an important question, non-evolutionary taxonomy would possess an important purpose.

That is the proposed justification. It only remains to discover whether it is valid. It would work with greatest possible power if two things are true: first, that classification is essential to test evolution; and second, that *cladistic* classification is essential to test evolution. We must therefore begin with three questions: Can classification test evolution at all? If it can, is classification necessary in order to test evolution? If it is, is it a *cladistic* classification that is necessary? Once we have answered them we can ask a fourth question, for none of these three directly challenges the proposed justification of transformed cladism. We must also ask whether it really is a circular argument to assume that cladism is phylogenetic, and also to use a cladistic classification to test evolution. We shall now take these four questions in turn. But, because the discussion will prove a little laborious, let me give, in advance, a four-word summary of the answers. If the justification of transformed cladism is valid, all four answers must be positive. The actual answers, however, taking the four questions in order, will be: Yes, No, No, No. Any one negative would destroy the justification, but three out of four, *cladis sunt tria partes*.

We must fix what I mean by a 'test' of evolution. A test can only operate between evolution and a specified list of alternatives. Before we can call something a test of evolution, we must specify what the alternative is. The most obvious alternative would be the biblical book of Genesis. But if taken literally, Genesis runs into trouble with some simple facts of geology. If I was asked why I do not accept Genesis, I should immediately seize on the geological evidence of the age of the Earth. That is exactly what the scientific party did at the recent trial at Arkansas, and it worked (Lewin 1982a,b). If we take some other alternative to evolution, a different test will probably be required. One other alternative is the theory of separate creation of the early editions of Lyell's *Principles of Geology* (e.g. 1830-3), according to which species originated separately, rather than by descent from a common ancestor. It need not be worried by the geological timescale. To refute it, we should need other arguments. As we consider how classification may test the theory of evolution, we must have some particular alternatives in mind. Let us here use these two, Genesis and the theory of separate creation.

Classification can argue for evolution against both of them. The hierarchical structure of taxonomy is often said to suggest evolution, and so it does, but the hierarchical structure alone is not sufficient proof. Any objects, whether or not they originated in an evolutionary process, can be classified hierarchically. Buildings, for example, do not physically reproduce; they have separate origins: but they can be classified hierarchically, by such criteria as whether they are large or small, ugly or beautiful, of brick or stone. Any particular set of buildings may not be divided hierarchically by these particular criteria; but an appropriate set of criteria should be discoverable for any case. This being so, the mere fact that biological classification is hierarchical does not prove that evolution has taken place. The true argument from classification is more subtle. It lies in the nature of the characters that define the classification. These characters have two important properties. One is that they are homologies, not analogies. The other is that they indicate natural, not artificial, groups. The two properties are closely related, for naturalness is used to recognize homologies (see p. 25), but it is convenient to take them separately. Homology, in a non-evolutionary context, refers to similarities between species that are not required by similarities in their modes of life: if fish and mammals had independent origins, the bones of the mammalian ear would not have to be the same bones as form part of the jaw in fish. In fact they are. If mammals had evolved from piscine ancestors the fact would be less surprising, because new parts must be formed by changes in preexisting parts. The homology, therefore, of ear bones in mammals and jaw bones in fish, is an argument for evolution. It argues against any alternative which postulates the separate origins of these two groups. Like ear bones, many other homologies are used to define taxonomic groups, and in every case the same argument applies. If classifications were defined by any old kind of character, rather than homologies in particular, the hierarchical nature of the classification would not be an argument for evolution.

The second point was that biological classifications are natural, not artificial. The members of taxonomic groups resemble one another for many other characters in addition to those that define them (examples will come later). This suggests evolution, because evolution generates natural groups (see pp. 19 and 58), but a hierarchical classification of non-evolutionary objects will usually be artificial: the members of the groups of ugly and beautiful buildings (for instance) will not resemble one another in other independent characteristics. The naturalness of biological classifications argues for evolution against any alternative which postulates an independent origin for the groups, at any level, in the classification. Naturalness and homology matter less because they are used in classification than because they suggest that the species are connected phylogenetically. The real test of evolution is the discovery of a phylogeny. It can, however, be called an argument from classification because the same evidence is used to classify species and to reveal that they are connected by a phylogeny. It does not matter whether the argument is called an argument from phylogeny or from classification; the argument itself is the same in either case.

Biological classifications, therefore, can test the theory of evolution. So far, so good for transformed cladism. Now let us turn to the stronger question of whether classification not only can test evolution, but is also essential to test it; of whether, in other words, classification is the only test of evolution. If it is, the justification of transformed cladism could (if the remaining steps in the argument succeed) turn out to apply in a strong form. If classification were the only test of evolution, evolutionary biologists would have to keep evolution out of classification; but if classification turns out to be only one among several arguments, then transformed cladism could only be justified in a weaker form. Evolutionary biologists would not have to exclude evolution from classification; and if we had any reason to include it, the justification of transformed cladism could be overruled.

The usual list of evidence for evolution includes several apparently non-taxonomic arguments. The crucial question is whether they genuinely are non-taxonomic, or whether they really reduce to the taxonomic argument we have just discussed. We can start with the geological argument, which Darwin discussed in two main forms. The first is the evidence of transitional stages between populations at different times. Darwin rejected this line of evidence, because of the incompleteness of the fossil record; and, despite the discovery of a few cases of fairly complete records, Darwin's argument still applies today. The second is the 'geological succession of organic beings'. The grand pattern of the fossil record, for instance in vertebrates in which fish arise first, then amphibians, then reptiles, then mammals, and finally the subgroups of mammals, is exactly what an evolutionist would predict. Maynard Smith (repeating a remark of Haldane, he says) once told a televisual audience that his evolutionary views would be severely disturbed if a fossil rabbit were discovered in the Precambrian. The fact that one has not is evidence for evolution. And not only is the geological succession predicted by evolution, it is decisively not predicted by Genesis. Lyell (1830–3), in his pre-evolutionary phase, was well aware of the geological succession, and tried to explain it by changes in the environment. But even if we allow his rather forced explanation, the geological succession argues against the biblical alternative to evolution. The argument is non-taxonomic. It is no use a transformed cladist saying that the argument assumes the animals have been classified (into mammals, primates, etc.), and reduces to the taxonomic argument, because it does not. I only used those names for convenience. The actual argument does not require any classification. The argument is that complex evolutionary changes must take place in a series of small changes by the modification of pre-existing parts: to get from a fish-like form to a primate-like form requires transition through reptilian forms.

The geological succession provides one non-taxonomic argument for evolution. It is not the only one. There is also the evidence of the artificial selection of evolutionary change – which has often produced new biological species – and there is 'ecological genetic' evidence of natural evolution. A transformed cladist might, in desperation now, plead that these only demonstrate microevolution. But that would be a feeble thing to say. All that is needed to prove evolution is observed microevolution added to the philosophical doctrine of uniformitarianism, which (in the form that is needed here) underlies all science.

That is not all. The evolutionist can, and still without resort to classification, point next to what may be called universal homologies. These are properties possessed in common by all species, but which would not have been if all species had been created separately. The genetic code is a universal homology. How do we know it is universal? Originally, because when Von Ehrenstein and Lipmann (1962) injected the mRNA for haemoglobin of rabbit into *Escherichia coli*, the bacterium then synthesized the protein; but this kind of experiment is now a biotechnological commonplace. Similarly, Weinstein and Schechter (1962; Weinstein 1963) found that a variety of different species all synthesized polyphenylalanine if injected with polyuridylic acid. So the code is universal; but how do we know it is

Evolution and Classification

homologous, and not convergent? The question (Crick 1966, p. 8) 'is whether the code has a structural basis, or whether it mainly arose by chance. If the former were true there should be some stereochemical relationship between each amino acid and the triplets that code for it, or with the corresponding anticodons. So far this has not been demonstrated.' Nor has it yet. The code is therefore thought to be a 'frozen accident' (Crick 1968), descended from a unique origin of life. There are other biochemical universals too: all amino acids are laevorotatory; and all species use much the same molecular genetic apparatus. They independently support the same theory. They are evidence for evolution.

Although the argument from universal homologies has the same form as the argument from classification, it is independent of it. We need no more than the most rudimentary classificatory knowledge in order to demonstrate the universality of the genetic code. Von Ehrenstein and Lipmann's experiment did appear to presuppose that *E. coli* is a different species from a rabbit. However, even if this degree of formal taxonomic knowledge were necessary, it should not comfort transformed cladism. The evolutionary philosophy repudiated by that school is used to classify the higher categories, not to distinguish species. Moreover, the demonstration would have been just as powerful with less formal taxonomic knowledge: all that matters is that the two forms in the experiment should be very different from each other. The injection of haemoglobin mRNA into lots of randomly chosen and unclassified forms could equally well demonstrate the universality of the genetic code. And just as the demonstration of universality does not depend on classification, nor does classification depend on the existence of universal homologies. Universal homologies are not used in classification. Taxonomists classify living species, and no taxonomic act depends on, or can affect, the existence of universal homologies. Although the arguments are similar, from the standpoint of taxonomic practice that is a coincidence. No taxonomist need be professionally aware of the existence of universal homologies. They are independent of classification. That, however, has not prevented them from providing yet more evidence of evolution.

What might we conclude at this stage? Classification can test evolution, but it is only one test among many. That would seem to open up two possible uses of the theory of evolution in classification. We could either use classification to test evolution, or we could test it by other means and then (if we wished) assume the truth of evolution in classifying. If we had no reason to use the theory of evolution in classification, transformed cladism might be able to find a weak justification as a test (though an unnecessary one) of evolution. If we had reason to use it – which in fact we do – then we could dispense with the classificatory test, and with transformed cladism too. But that would only be the conclusion if the remaining two questions receive positive answers. Let us turn to them now.

We have demonstrated that classification can test evolution. Cladism therefore can as well (Cracraft 1983). That, however, does not add up to a justification of transformed cladism. It must be shown not only that cladism can test evolution, but also that the techniques of numerical and evolutionary taxonomy, if applied without presupposing evolution, could not. For the techniques of any school can be operated without assuming evolution. The technical acts of the taxonomist have no need of the assumptions that justify the techniques. Of course, it would be rather absurd to distinguish 'convergent' from 'ancestral' similarity in the evolutionary school if evolution were untrue, but it is exactly that kind of absurdity, in cladism, which is under analysis. We cannot ignore it. Transformed cladists make the comparable distinction between ancestral and derived characters without an evolutionary assumption, and evolutionary taxonomy could do likewise. The technical deed could be done. And because it could be done, if it is not going to be it must be for a reason, not because it is practically impossible. Any school that would justify itself as a test of evolution must show that it alone can provide such a test. If it does not, it is unjustified: we should be left at most with a justification of classification in general, and a subjective choice among particular schools. If transformed cladism is to be justified as a test of evolution, the taxonomic argument for evolution must work only with cladistic techniques. Let us see whether it does.

To state the taxonomic argument should be enough to settle the matter. It is (as we have seen) that the classificatory hierarchy is natural, not artificial, and is defined by homologies. Now, cladistic classifications are natural, but then so too are numerical and evolutionary classifications. Likewise, although cladistic classifications do use homologies, so too do evolutionary classifications. The need to test evolution could therefore at best only justify all taxonomic schools equally, and none of them in particular. It does not justify transformed cladism.

The same point can be made by example. E. B. Ford (1941, 1942, 1944a,b, 1947) pioneered the use of chemical characters in classification; but he also used them to test whether pre-existing classifications were natural. He worked on Lepidoptera. Pigment chemistry, of course, had not originally been used to classify the Lepidoptera. Taking specimens from the recognized families, Ford discovered whether the distribution of pigment chemicals matched the accepted classification. It turned out that they did. Such would have been the prediction of evolution, not of separate creation. But it did not require transformed cladism! Ford did not distinguish ancestral from derived states of pigment chemistry, nor classify by shared derived

Evolution and Classification

pigments. He supposed that similar pigments were homologous in the evolutionary taxonomist's sense (see p. 30). A similar kind of test has been performed more recently, although (now that the creationists and Popperians are among us) it was divulged differently. It also used a chemical character, protein sequences. Penny, Foulds, and Hendy (1982) reasoned that, if evolution is true, the pattern of sequence differences among species should be the same for different proteins. A taxonomy formed by any one of them, in other words, should be natural. They accordingly compared the tree of 11 species, formed from the sequences of five different proteins. The five trees were topologically identical: the prediction of evolution, not of separate creation was confirmed. But although Penny, Foulds, and Hendy thus offered evidence for evolution, and by a taxonomic argument, they (like Ford) did not use cladistic techniques. They did not distinguish ancestral from derived amino acid similarities. They supposed only that similar protein sequences were homologous in the evolutionary taxonomist's sense.

The end has now come for transformed cladism. It is not the only taxonomic school that can test evolution: any of the main schools can. The need to test evolution (which is real but which need not be met by classification) could justify any of the main schools. We could stop here, but I believe it is worth examining the fourth question too: not to turn the knife in the wound, but because it is of general interest. Let us admit therefore that classification is needed to test evolution (which it is not), and that the kind of classification needed must be constructed with cladistic techniques (which it need not). Then we can ask whether our philosophy of cladism must be Hennigian, which assumes evolution, or transformed, which does not. Is it circular to argue that classification tests evolution, if the truth of evolution was assumed in building the classification?

We have met this kind of question twice before. Numerical taxonomists criticized the evolutionary taxonomist's method of recognizing homologies because (they said) it was circular (see p. 27). A similar charge has been made against the cladistic technique of outgroup comparison (p. 64). Actually, it is only a small step from the numerical taxonomic critique of how homologies are detected to the argument of transformed cladism. Numerical taxonomists, as it happened, fired their critiques against the methods by which homologies were recognized; but then, in the 1950s and 60s, you could scandalize Darwinian society more easily if you professed that the course of evolution could not be discovered, than that it had not taken place. But the step from one argument to the other is small, and it is not surprising that it was taken in one of those numerical taxonomic works:

> [Darwin] did not commit the earlier error of arguing in a circle. His principles of evolutionary importance were not

derived from a pre-existing taxonomy, but from the results of artificial selection and from the study of heredity, variation and ecology. This is a point worth emphasizing. In many elementary text-books of biology, classification is treated as one of the lines of evidence for evolution. Darwin did not treat it thus; he discussed it quite late in the '*Origin*' as consonant with the theory of evolution, and explicable as a consequence of it. He never regarded it as primary evidence for evolution, and his caution was certainly justified. (Cain 1962, pp. 9-10)

We have here an argument very like that of transformed cladism. And just as the arguments are very similar, so the same defence will serve. It would indeed be possible to formulate the argument as a circular one, and perhaps some authors have mistakenly done so. However, the circular formulation is the argument in its worst form, not its best, and we should be interested in the latter. The argument properly proceeds not in a circle, but by 'successive approximation'. The truth of a general proposition must first be independently confirmed, and then when this has been done, it can be used as an assumption in interpreting later observations; if the interpretation is successful, our confidence in the theory is proportionally increased.

First, an independent test must confirm the truth of evolution. This could be achieved by any of the non-classificatory arguments for evolution. Or we could classify a small group of animals without the assumption of evolution, and test the theory with classificatory evidence. We might, for instance, test the theory of evolution on the birds. It would be important not to assume the truth of evolution while doing this, as it is an independent test. If the test was successful we could then, if we wished, accept the truth of evolution in future classifications. As each natural classification, and each homologous similarity, was discovered, the evidence for evolution would increase. We might, in other words, have needed to be transformed cladists for a short period last century, while classifying some small group; but the need has now passed. Now evolution has been confirmed, we can, without any circular reasoning, use the theory to build our classifications. In practice, the process of successive approximation is not as simple as that. In a logical reconstruction there must be an independent test, before subsequent assumption; but history need not proceed logically (Hull 1980). There need be no clear first stage in which the hypothesis is tested, and second stage during which it is assumed to be true. The real process is a continual feedback from the results of successive new tests, to our understanding of the facts in question.

The sequence – hypothesis, test, further hypothesis (assuming the results of the previous test), further test – is not circular. Nor is it scientifically peculiar. What the numerical taxonomists, and now transformed cladists, really object to is not circular argument, but theoretical ideas. Theories in science, if they are interesting, are always taken beyond the evidence, and assumed in further tests. The machinery of molecular genetics has only actually been confirmed in a very few species, but biologists unhesitatingly agree that it is universal. They use this knowledge, obtained in *E. Coli*, in interpreting the results of experiments on all other species. We can test Newton's law of gravity with an apple in an orchard, and then apply it to the motions of the heavenly bodies; we can likewise test the theory of evolution in a genus of fruitflies, and apply it to all of life.

We therefore can both assume, and test, evolution in the process of classification. The only question left open by the argument of this chapter is whether we ought to. We have been through all that before. In Chapters 2-4 we examined the possible taxonomic systems, and argued that the best system is phylogenetic. The best known philosophy of classifications, therefore, depends on the theory of evolution.

Taxonomic controversialists seem fatally attracted to the discovery of circular arguments in the systems of their opponents. It is not only transformed cladists who have found circles in evolutionary systems: numerical taxonomists have found them in evolutionary and in cladistic taxonomy (see pp. 27 and 85), evolutionary taxonomists have found them in phenetic taxonomy (Johnson 1970), and in cladism (Bock 1981); some cladistic factions have found them in the systems of other factions (Schaeffer, Hecht, and Eldredge 1972), and I do not suppose the list ends there. The remarkable point is that they have nearly all been wrong. It is all too easy to see circles where there are none; and, with this history, my suspicions are always raised when I hear the charge being made.

The conclusion of this chapter is that transformed cladism cannot be justified as a test of evolution. The justification contains three lethal errors. Although classification can provide a test of evolution, it is not a necessary test. Even if it were necessary, this would not justify transformed cladism, as opposed to any other technique of classification. Finally, there is nothing wrong, or circular, in both assuming evolution in order to classify, and using the classifications to prove evolution.

We have how examined all the proposed justifications of transformed cladism, and can draw a common conclusion. None of them work. Transformed cladism is either nakedly subjective, or can easily be stripped of its thinly deceptive justifications, and exposed in its truly subjective form. There is a good reason to use cladisitc techniques, but it secures them by logic to the theory of evolution. If the theory of evolution is abandoned, so should be cladism.

8

The functional criterion

Although the rejection of evolution from classification is the most far-reaching ambition of transformed cladism, the same revolutionary movement has made several other controversial demands. Indeed, within taxonomy, the rejection of evolution has probably been discussed less than three other narrower practical issues. The main methods of classical 'evolutionary taxonomy' (Ch. 2), at least according to its practitioners, are functional analysis, to distinguish homologies from analogies, and palaeontology, to recognize ancestral forms: but many cladists have recently denied that either method works, and have called for their abolition. A third cladistic demand has also proved highly controversial, which is that classification should preferentially be dichotomous. In the next three chapters we shall consider these three controversies in turn.

The concepts of function, adaptation, and natural selection all go hand in hand. The 'function' of a part of an organism is the way in which it is adapted; to understand how an organ functions is to understand how it is adapted, and why natural selection brought it into existence. We have encountered two proposed classificatory uses for this set of concepts. The grandest was the possibility of a whole teleological principle of classification, in which species would be grouped according to their principal adaptations; but I have no more to say about that. The other, which we met in evolutionary taxonomy, was to distinguish homologies from analogies. This we shall consider further in this chapter, together with a third possible use, to distinguish ancestral from derived character states.

For now we have decided in favour of the Hennigian school, the outstanding problem of classification is to recognize accurately ancestral and derived characters. We do possess some methods (as we have seen) but they are far from perfect, and we could well do with some more. Moreover, because our existing methods often mistake shared derived characters, we also need methods of reconciling apparent contradictions among characters. In the absence of

Evolution and Classification

functional evidence, the main such method is the principle of parsimony. According to that principle, in its simplest form, the characters that fit the majority pattern have been correctly identified, and are homologous; those in the minority have been mistaken, and are really different characters that have converged to look the same. Here we shall consider how the principle might be made more sophisticated. In its simple form it counts each character equally; but if some kinds of character are more liable to convergence, they could be weighted less in estimating which is the 'majority' pattern. Functional analysis is, of course, the main method that might reveal which characters are liable to convergence. We shall concentrate on how it might be applied to the twin problems of distinguishing ancestral from derived states, and distinguishing truly homologous shared derived characters from analogies (instances of apparently shared derived characters that are really different characters). The problems are closely connected.

Neither of these applications could assist purely phenetic classification. In that school, species are clustered despite conflicts of characters, and regardless of the distinction of ancestral and derived states. Characters are indiscriminately selected and measured: there is no use for any theory of character analysis. Numerical taxonomists such as Cain did realize that natural selection might provide a technique of character weighting, and discussed the possibility (Cain ed. 1959b; Cain and Harrison 1960); it could only have become of interest if numerical taxonomy had developed from a phenetic into an evolutionary (or even teleological) system, but in any case it never became a practical method. Evolutionary taxonomy might desire to distinguish ancestral from derived character states, but that school, as I have identified it, classifies groups by shared homologies, and is more interested in the functional distinction of homologies from analogies. Of the three schools, only cladism keenly needs both applications.

Notwithstanding this conceptual requirement, cladists have almost universally rejected functional analysis; to argue for its use from within cladism is at present a peculiar, if not unique (Fisher 1981), position. Its offence is to do violence to the 'pattern-process' distinction: natural selection is a theory of process; if, therefore, it could be used to investigate classificatory patterns, the study of pattern and process could not be kept separate, contrary to the doctrine of transformed cladism. Indeed, part of the interest of considering functional methods is to test the transformed cladist's distinction of pattern and process. If the functional criterion can distinguish ancestral from derived character states, and homologies from analogies, it should be admitted as a cladistic technique, and the pattern-process distinction will be undermined. If the criterion is to be rejected, and the distinction saved, it must be because argument

The functional criterion

has shown the criterion to be in error. These are the stakes: all we need now are the arguments. Unfortunately the literature of cladism is little help here. The exclusion of functional analysis has been accomplished mainly by silently overlooking it, or by scattered critical asides, and not by detailed argument. In practice, and in recounting their principles, cladists have preferred to distinguish ancestral from derived character states by methods such as embryology or outgroup comparison, and to reconcile character conflicts by naive parsimony. The silence, however, is not absolute, and one author (Cracraft 1981) has written his objections down in a whole paper, where we can examine them. The paper is a polemical reply to a polemic by Bock (1981). We most allow for some exaggeration in exchanges of this sort, but Cracraft's stated conclusion is clear enough. There is no 'strong case for the need for' (p. 30) functional analysis of characters; it is not a 'necessary component' (p. 35) of cladistic analysis. Bock had argued that functional analysis is necessary and essential to classification; Cracraft replied that it was not. Now, although I happen to agree with Cracraft, this chapter will mainly disagree with him. I suspect there is a stronger practical conclusion behind the stated one that functional analysis is unnecessary: namely, that functional analysis is impossible, and should be left alone. the thesis I shall defend is not that the functional analysis of character states is necessary: it is only that it is possible. There is a weaker sense in which it may be said to be necessary as well, which I shall notice at the end – it is that because it is difficult to distinguish ancestral from derived character states, we do well to use all possible kinds of evidence.

What, then, does Cracraft object to in the functional criterion? His main arguments do not actually concern the question of whether natural selection can be used to determine the polarity of character states. He is concerned more with the concept of adaptation itself, presumably because if you cannot recognize what an adaptation is to begin with, you certainly will not be able to use any further argument in which the concept of adaptation is essential. According to Cracraft, the criterion of an adaptation is subjective and authoritarian (pp. 26, 29): it is a 'belief system relying primarily on an appeal to authority' (p. 30). Humphries (1983) would agree. He describes the functional criterion of homology as that 'alchemy of character weighting so inherent to "traditional" authoritarianism we try so hard to escape' (p. 308, also p. 305). We might just note in passing that the subjectivity, or authoritarianism, of the concept would not be an objection to its use within transformed cladism. In that system, as we have seen, justification is not considered necessary if only a method can be applied. As long as we relied on a consistent authority, the functional criterion, even if Cracraft were right, could be admitted directly into transformed cladism.

Evolution and Classification

But is he right? Is the science of adaptation subjective and authoritarian? Like some other critics of the science, Cracraft does not credit it with its most powerful form. It is not enough only to quote poor work, as if it was representative; all sciences have their poorer sides. There is rigorous work on adaptation too, and it is in this that he should have looked for subjectivity and authoritarianism. Consider the experiments of Kettlewell, and other ecological geneticists (Ford 1975), of Lack (1966) on clutch size, the work on sex ratios (Charnov 1982) and evolutionarily stable strategies (Maynard Smith 1982): are these subjective and authoritarian? Cracraft does not mention one of them. He (p. 31) actually believes that nothing can properly be called an adaptation until it has been shown to vary intraspecifically, and that the variation is heritable. He is wrong (see, for example, Dawkins 1982, Ch. 2, especially pp. 20-1, 26). He thinks that the concept of adaptation cannot be applied at all to interspecific comparisons; but interspecific comparison has always provided powerful evidence of adaptation (Ridley 1983).

Cracraft's general objection does not work. But if the criterion is to be used it is not enough to defend it successfully against its critics: we must also justify it by positive argument. Let us therefore turn to its two proposed applications, and see whether they are subjective belief systems, or rational scientific methods.

The functional criterion of homology

The problem is to distinguish homologies from analogies. Analogies are due to convergence; and convergence is due to natural selection: if, therefore, a character shared by two species is likely to be convergent, it is unlikely to be homologous, and vice versa. We saw some applications of the method in Chapters 2 and 5 (to the case of 'loss convergences'). Those discussions can here be counted in favour of the method, for they suggest it can be used. So too can the work of Fisher (1981), who applied the method to character conflicts among species of *Limulus* (he also provided a similar discussion to mine). A slightly different application of the same method can be found in the work of Manton (1977). Her interest was the polyphyly of the characters of arthropods. Her method was to compare organs, which were thought to be homologous, in different groups, to see whether they were exactly the same; if they differed she then asked whether the intermediates between the forms would have been functional arthropods. Her analysis of limb joints (summarized in 1977, her fig. 5.15) provides an example. The limbs of the different insectan groups, and of the myriapods, articulate on the body differently. One form (she tells us) could not have evolved gradually into another if all the intermediates had hard exoskeletons. The intermediates (she reasoned) were probably soft-bodied, and the different myriapodan groups and the insects probably arthropodized independently from a soft-bodied ancestor. Two points of clarification may be necessary. Jointed limbs and hard exoskeletans are distinctive arthropodan characters; but not all arthropods, especially larval stages, possess them. All these cases, however, can be interpreted as descended from an ancestor that at some stage of its life had a hard exoskeleton: we might call the soft exoskeleton a 'modified' exoskeleton, using the same trick to confound pedants as was used in a footnote on page 96 above. Secondly, Manton's method applies to characters rather than groups. Her argument may show that the ancestor of insects and myriapods did not have a hard exoskeleton; that, however, does not in itself mean that they do not share a more recent common ancestor with each other than with any other arthropodan group. The ancestor might have been soft-bodied. The shared derived arthropodan characters of insects and myriapods had previously been thought to show that they formed a monophyletic group (Figure 8.1), descended from an ancestor with a hard exoskeleton. Manton's analysis, of this character at least, points to a different conclusion. It suggests the shared derived character is not a shared character at all, but was evolved independently in each. The relations of the three groups

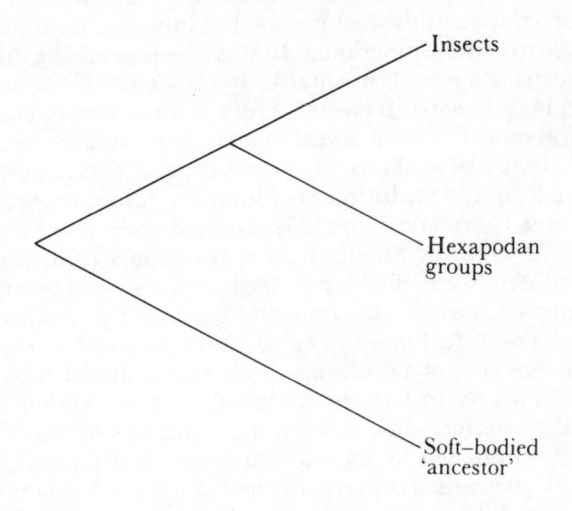

Figure 8.1
Pre-Manton classification of insects, myriapods, and hypothetical soft-bodied ancestor.

Evolution and Classification

(insects, myriapods, soft-bodied ancestor) therefore become undetermined. (By a similar argument, but for jaw mechanisms, Manton intruded the soft-bodied Onychophora as the sister group of the insects and myriapods, to leave the hard-bodied Crustacea and Arachnida as more distant relatives.) The conclusion is not proved simply by showing that the character states are different in the insectan and in the myriapodan groups; which is necessary, but unremarkable, because no characters in any two organisms, let alone two classes, are identical. The important part of the proof, as Manton insists, is that the intermediates between the groups could not be functional and arthropodan. The limb joints are not ancestral.

Manton's conclusions are controversial, but that is not the issue here. What matters is whether her method is in principle valid. The answer is, that it is. It proceeds, in summary, as follows. Take a supposedly shared derived character in two groups. Consider whether the intermediates between them could have been functional without the character changing back to some ancestral state: if they could, the hypothesis of shared derived character is confirmed; if they could not, it is refuted; if we cannot be sure, the conclusion for the character is uncertain. The method is applicable to all possible homologies. It applies to shared ancestral characters as well as shared derived ones: one could, for example, use it to assess whether a supposedly shared ancestral character such as the backbone of fish and mammals was homologous. For any intermediate it is perfectly possible to ask the strong question of whether it would be functional at all, and the weaker one of whether it would be more or less functional relative to the end points. If a limb could not walk, it is determined to be non-functional. In this case it would be difficult to test the capability experimentally, but that does not mean the assessment is subjective: it is not: it is derived not from fiat but from a mechanical model. A mechanical model can rationally demonstrate whether a limb can walk. In other cases experimental intermediates can be made; in the evolution of coloration, for example (Schmidt 1960; Brower, Cook, and Croze 1967; Cook, Brower, and Alcock 1969; Benson 1972; Teffords, Sternburg, and Waldbauer 1979; Silberglied, Aiello, and Windsor 1980). Many more examples could be cited, but it is my purpose only to illustrate the form and possibility of the technique, not its full range of application. In all cases, the general argument has one of two forms: a character shared between two species is unlikely to have been shared from a common ancestor either if the intermediates between the characters of the two would have been eliminated by natural selection, or if natural selection might well have caused convergence between them. In many cases the conclusion will be uncertain, but even this can be valuable positive knowledge. When we are seeking to reconcile conflicting characters, it may help to know that some of the shared characters are uncertain.

The functional distinction of ancestral from derived character states

Can we, from a knowledge of adaptation, distinguish ancestral from derived character states? Cracraft (1981) concluded in the negative, after directing his criticisms against one particular method. That method was first to arrange the character states into a sequence of small changes; and then to assess the evolutionary polarity of the sequence by the principle that natural selection must direct evolution towards an increasing, not a decreasing, degree of adaptation. The least well-adapted state would then be ancestral, and the successively better adapted states successively more derived. The method is a classical argument of evolutionary morphology, discussed by Severtzoff (1929), and now recommended by Bock (e.g. 1981), Gutmann (e.g. 1981), and others. I suspect it has been more often discussed than used. It has a curiously strong influence in theoretical discussions, where it may pass for the only functional criterion (Cracraft 1981), or even for the only cladistic technique (Foelix 1982). It has, however, been used sometimes; the references above contain some examples, and Fisher (1981, p. 56) has applied it to *Limulus*.

Bock and Gutmann have spoken for themselves, and I do not intend to say much more about the technique. Maybe it is sound; and maybe it is not. It is unfortunately tied to the theory of evolutionary 'progress', a theory which, I believe, cannot satisfactorily explain organic diversity. Character states may differ, not so much because they are more or less well adapted to the same niche, but because they are adaptations to different niches. However, Bock and Gutmann's functional criterion is just one functional method among several. Even if their form of it were refuted, that would not mean that the whole functional criterion should be cast out of cladism. I shall concentrate on two other, non-progressionist techniques.

Let us start with the case of vestigial organs. Fish, shrimps, and isopods that inhabit underground caves possess organs in many respects similar to eyes; in various degrees they have lost parts, such as pigments, that are essential if the eye is to function (Culver 1982). They are more or less blind. The internal remnants of limb bones in whales are a similar functionless vestige. Darwinians can be as certain as they are of anything that the vestigial eyes of cave populations of the isopod *Asellus aquaticus* are a derived character state relative to surface populations with functioning eyes, and the hind-limb bones of whales relative to those of other mammals. The judgement is supported by the functional criterion. A complex adaptation such as an eye must be built up in many small stages, each of them functional. If, after it has been built up, selection favours the

Evolution and Classification

loss of vision, the evolutionary course of loss may differ from that of its initial evolution. Figure 8.2 shows the direction of evolution. The vestigial organs can be recognized because the course of loss differs from that of initial evolution. They are a derived character state. The functional criterion dictates that they could not be ancestral with respect to the full eye. The criterion is that a non-functioning version of a complex adaptation must be a derived state. If the stages in the loss of an organ are exactly the same as those of its initial evolution then the criterion cannot be used.

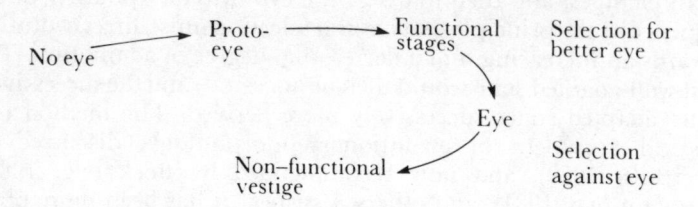

Figure 8.2
The course of evolution, and decay, of a complex organ, the eye.

The useless eyes of cave animals are only an extreme instance of a more general phenomenon. Many functional characters contain, within their structure, subtler vestiges of their history. Le Gros Clark (1959) discusses several examples. The canine tooth of humans, for instance, is a functioning organ, but possesses vestigial information. It is small compared with that of other apes (Figure 8.3) but not compared with many other mammals. Is the small canine of humans the ancestral condition for apes, retained from other mammals? Or is it secondarily derived from a larger ape-like condition? Here is how Le Gros Clark answers the question (1959, p. 6):

> The canine tooth in *Homo sapiens* is a small, spatulate tooth which shows very little differentiation in comparison with the adjacent teeth. It has actually been argued that this is more primitive than the condition in the anthropoid apes in which the canines are large, projecting, and pointed teeth. But, as we shall see, the real nature of the human canine as a tooth which has secondarily undergone some degree of reduction is betrayed by certain characters such as the length and stoutness of its root (which seems out of proportion to the functions it performs), its relative lateness of eruption, and the fact that the newly erupted tooth may occasionally be sharply pointed and somewhat projecting.

The human canine therefore appears to possess vestiges of the structure of a much larger tooth, which suggests that its reduced size

The functional criterion

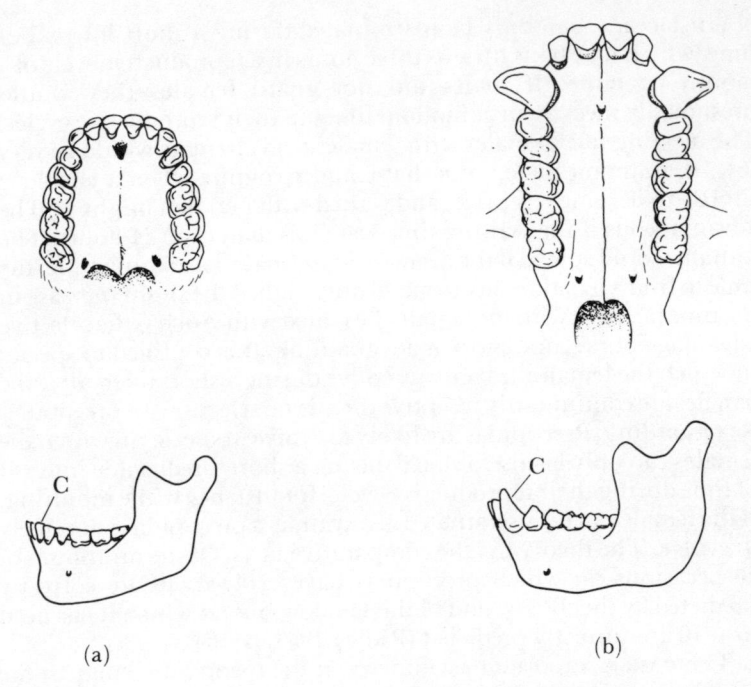

(a)   (b)

Figure 8.3
Upper and lower dentition of (a) human and (b) great ape (actually, the lower dentition is of a male orang-utan and the upper dentition is of a male gorilla). Note relative sizes of canine (C) in human and ape. (Modified from Le Gros Clark, W. (1959), *The Antecedents of Man*, with permission from The University of Edinburgh Press.)

is a derived condition. Le Gros Clark offers analogous arguments concerning the polarity of the evolution from claw to nail (pp. 6, 173–4; see also Szalay 1981), and the simian sulcus of the braincase of monkeys and apes and its apparent absence in humans (pp. 6–7). Again, my main point is that the method exists, is rational, and is practical.

The cladistic distinction can be made even without vestigial information. All that is needed is an asymmetry of the direction in which natural selection would drive evolution (Ridley 1983, pp. 34–40). This is little explored territory, and I know of only fairly obscure examples. In comparative studies of adaptation, some transitions between character states are often relatively less likely than others. For instance, I have surveyed the incidence of precopulatory mate guarding (Ridley 1983), a reproductive habit in which the male stays very close to the female for a lengthy time before insemination. The theory is this. Females in some species, particularly of crustaceans, can only be inseminated during a short interval of time, which may be as little as three hours in a reproductive cycle of a month or more. If males did not guard females they would presumably meet them at random times in their reproductive cycles. The meeting rate of males with females ready to mate would be very low. If a mutant male arose that could recognize when a female is shortly to become receptive, and waited with her, that might well be advantageous if the waiting time were less than (say) 24 hours. The mutant would spread if the decrease in the male's searching time (the time to find a female ready to mate) outweighed the slight increase in the time he spent with the females he mated with. Such is the selective advantage of precopulatory mate guarding. It is confined to species in which the females are only receptive during a short interval: if the females are continuously receptive there is no selection for precopulatory guarding. Precopulas are likely to evolve in species in which the females can only be inseminated during a short, predictable interval of time during their reproductive cycle, for instance while moulting. If the female can be inseminated at any time, a precopula is not likely to evolve. The theory fits the comparative facts. Of the minimum of 20 occasions on which precopulas have evolved, 19 are correctly predicted by the theory; and of the 11 occasions on which it has been lost, 10 are correctly predicted (Ridley 1983, p. 164).

There is an important asymmetry in the theory. A change in the female cycle is likely to lead to a change in whether or not there is precopula; but there is no reason, in the theory, why a change in whether there is a precopula should lead to a change in the female reproductive cycle. And the asymmetry can be put to cladistic use. We have two characters, each with two states: precopula (p) or no precopula (n), and insemination confined in time (c) or unconfined (u). The theory states that pc and nu are adaptive but pu and nc are not. It also states that some transitions are more likely than others. We can write down four kinds of species, pc, pu, nc, and nu, and ask for each what kind of species it may and may not evolve into. The exact analysis might take more than one form, and the theory does not make clear predictions for all the transitions; but a preliminary analysis is possible. If species are pu or nc they fall outside the theory. They may be non-adaptive, and about to evolve into nu or pc, respectively (this is Bock and Gutmann's method); or they may be adaptive for some reason that the theory does not yet take into account, in which case all we could say is that they will not evolve into the complementary 'non-adaptive' state. We can deal with the two adaptive states more confidently. A species with a precopula and mating confined in time (pu) would not (if the theory is correct) evolve into either of the two non-adaptive states pc or nu. The theoretical transitions are in Table 8.1. I have left out some transitions for which the prediction is either uncertain or not

The functional criterion

Table 8.1. Probable (y) and improbable (n) transitions among the states of two characters, presence (p) and absence (n) of precopula, and mating confined (c) and unconfined (u) in time in the female reproductive cycle.

		Derived species state			
		pc	pu	nc	nu
Ancestral	pc	(y)	n	n	
Species	pu	y	(y)	n	y
State	nc	y	n	(y)	y
	nu		n	n	(y)

The assessments are given for transitions from the state in the vertical column to that in the horizontal row. Parentheses indicate the assessment is trivial.

immediately obvious. We can read off the polarities of the character states from the table. For example, in a group of species with *pu* and *pc*, *pc* is probably the ancestral state, because *pc* will not evolve into *pu*, but *pu* will evolve into *pc*. (The adaptive complexes of characters in this example look slightly different from the states of single characters, such as a, a', familiar from the previous discussion of character polarity. The difference is purely superficial. It is a trivial transliteration to re-write the four characters as four states of one character: a_1, a_2, a_3, a_4. The same argument applies exactly.)

The improbable transitions (written n in the table) make it possible to use the functional criterion to judge the polarity of character states. To return to Cracraft's objection, it is perfectly straightforward to ask whether species with a confined interval of insemination will evolve a precopula, and whether species with a precopula will tend to evolve a confined interval of mating. The theory is not authoritarian; it has been made public and is open and accessible to all. It is not subjective; the consequence follows from the assumptions by the necessity of reason, not because anyone chose that it should.

The general form of the method is to consider whether one direction of evolution is easier than its opposite. The course of evolution of vestigial organs is such that they can evolve more easily from complex adaptations than complex adaptations from them; the evolution of a precopula after a change in the reproductive cycle is easier than is a change in the cycle after a change in the precopulatory mating habit. Dollo's law, according to which evolution is irreversible, is another version of the same general principle. For some characters, both directions of evolution may be equally likely. In those cases, the functional criterion cannot determine the direction of evolution. If the table of transitions (like Table 8.1) for a character permits all changes, the functional criterion will not help

Evolution and Classification

the cladist. Like any other method, it can be applied to some characters, but not to others. In the same manner, the fossil record cannot be applied to characters that lack a fossil record, the embryological criterion to characters that lack an ontogeny, and outgroup comparison to characters that are represented in all states in the outgroup.

The functional criterion will not usually give certainty. It is virtually certain in the case of vestigial organs, but they are an exception. Usually, the method will not allow us to conclude anything more than that one direction of evolution is more probable than the other. No doubt the further study of any particular case might often make the conclusion more certain, but at any one moment we may have to live with uncertainty. This will not please Cracraft (1981). He objects strongly to the probabilistic conclusions of the functional criterion: 'Along with Zuri and Bock other authors have stated their belief that functional criteria are useful in discerning homology from convergence (e.g. [four refs]), and all seem concerned with establishing the *probability* level of some shared similarity being homologous' (pp. 26–7). And he then asks 'how can one assess that probability level other than by an appeal to authority?' We have already answered him. Neither of our two formal techniques is authoritarian, and both can establish which is the more likely of two directions of evolution. To begin with, that is all we need. We wish to know which of the states is more likely to be ancestral; we do not actually have to know whether the relative probabilities are 100% and 0%, or 51% and 49%. More quantitative estimates could be made; but they are not necessary in a first analysis. The functional criterion does indeed give more certain results in some cases than in others: and these estimates might be of use in sorting out non-congruencies among characters. The degree of certainty could be expressed by ranking, and tentatively incorporated into a likelihood technique of reconstructing cladograms.

Uncertainty, of course, is not unique to the functional criterion. It is found in all science, even all rational inquiry. As we have seen in an earlier chapter, all the cladistic techniques are uncertain. Outgroup comparison has difficulties with convergence, the embryological criterion with neoteny, and the palaeontology with the incompleteness of the fossil record. Any technique that does not use a time machine will be uncertain. If the functional criterion only gives probabilities, so too do all practical criteria. When a criterion is restated as a formal technique, it may appear to be giving certain conclusions; but the appearance is completely illusory. The certainty is no more than an affectation. And if it is the affectation of certainty which anyone wants before admitting the functional criterion into cladism then just let him listen to the confident tone in which Manton (1977) discusses her own work:

The functional criterion

> The newer functional approach to arthropodan structure is yielding results of the greatest importance. This type of work, using induction based on reliable evidence, is replacing the older speculations which were made because the fossil record provided no evidence. Large edifices of theory, which have been raised on non-functional bases, can now be set aside ... The older theories are untenable because they imply other ancestral animals which could not function, or which did not exhibit progressive evolution of the right kind. (p. 35; see similarly pp. 36, 224, 225, 281, etc.)

And so on. Her style is well-enough known. Few share her certainty in her conclusions. But we do not need absolute certainty. The point is that we have a question which we want to answer: How can we discover the direction of evolution? It is difficult enough to answer without making it more difficult by throwing away possible lines of evidence. We should attack the question with all the means at our disposal. By putting together several lines of evidence, all of them no more than probable, we shall arrive at conclusions more certain than if we artificially limit ourselves to fewer lines of evidence. No single line of evidence is necessary, but that is not a reason to ignore any of them. The question, in a field like this, is not whether a method gives us certain conclusions. It is whether it gives us any information at all. If it does it should be used, and thought should be given to its improvement. The functional criterion should therefore be admitted to the repertory of cladistic techniques.

9

Ancestral groups and sister groups

If two species resemble each other more closely than any other species, they may bear one of two kinds of phylogenetic relationship to each other. They may either be sister species, or a pair of ancestor and descendant. The sister group is that other group with which a taxonomic group shares a more recent common ancestor than it does with any other group; the ancestral group is the one that it was derived from during evolution. The ancestor–descendant relationship is only possible between species at least one of which is a fossil. This is not only because an ancestor must precede its descendant in time: a stronger cladistic reason also rules out ancestor – descendant relations for modern forms. When a descendant species evolves from an ancestor, the ancestor may suffer any of three fates (Figure 9.1). Two present no problem: the ancestor may go extinct, in which case it will not be found among modern forms (Figure 9.1a); or it may itself evolve into some new, different-looking species, which will then be the sister species of the descendant (Figure 9.1b). But what if the ancestral species persists unchanged alongside its descendant (Figure 9.1c)? In an extreme case, the gene frequencies of the ancestral species might be unchanged after the descendant split off. How should the ancestor be classified after it has split? A phenetic classification would identify A – S and D of Figure 9.1c and, if a modern sample contained both species, it would then interpret them evolutionarily as ancestor and descendant. But that is not possible in cladism. However close the phenetic similarity of A and S they are genealogically different kinds of entities: in cladism, species are classified not by their phenetic similarity but by their position in the phylogenetic hierarchy; Figure 9.1b and 9.1c therefore represent cladistically analogous events. In both cases S is a different species from A, and the only possible two species sample of modern forms (S and D) contains a pair of sister species. Therefore, even when an ancestral form remains unchanged alongside its descendant, it is not, in cladism, recognized as the ancestral species.

Ancestral groups and sister groups

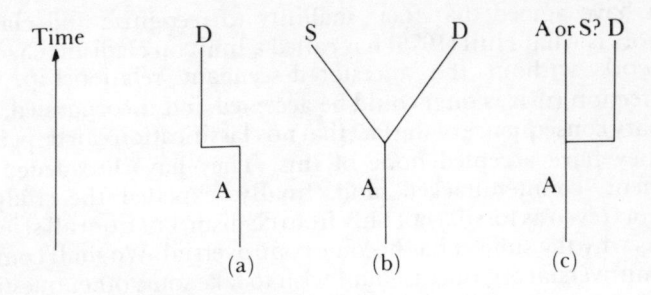

Figure 9.1
An ancestral species may suffer any of three evolutionary fates (A = ancestor, D = Descendant, S = sister species): (a) ancestral species goes extinct as its descendant arises; (b) ancestor evolves into two new forms; (c) ancestor persists alongside its descendant. Phenetic classification may distinguish cases (b) and (c); phylogenetic classification (cladism) does not.

So ancestor–descendant pairs cannot be found among modern forms. But what about fossils? A fossil sample could still contain species bearing either kind of relationship. Both kinds of relationship are real: they do both exist in nature. But the fact that they are real implies nothing about how they should be represented in a classification, and evolutionary and cladistic taxonomy treat ancestors quite differently. Evolutionary taxonomy comprehends the distinction, and at least tries to recognize ancestors. Cladism, by contrast, whether the species are fossil or modern, always confounds ancestors with sister species, and appears to classify them both as sister species. This cladistic procedure has proved controversial. The controversy arose out of what may be called the 'obvious interpretation' of cladism, that it means what it says: when it classified species as sister species, it means that they are sister species; when it excludes ancestors from cladograms, it means that no ancestors have been found. Given this interpretation, evolutionary taxonomists have not surprisingly claimed an advantage for themselves. Cladists represent all relationships as sister groupings, but is it so much more likely that closely related species are sister species than ancestor and descendant? Harper (1976, p. 184), for one, doubts it. He wrote, in a palaeontological paper: 'given two closely related taxa A and B, I regard the hypotheses A→B and B→A as, a priori, as plausible as the hypothesis that they originated from a common ancestral taxon not represented in the fossil sample studied'. (Szalay 1977 and several of the authors cited on page 32 make similar criticisms.) Because every species does have an ancestor, closely similar species could possibly be an ancestor and its descendant. Cladistic classification, however, appears to deny this.

The cladistic school might have accepted this criticism. They

139

Evolution and Classification

might have agreed that their inability to recognize and classify ancestors is what Hull (1979) has called a limit on cladism. Cladism can work without the ancestor-descendant relationship. The imperfection (if it is one) could be accepted and incorporated, as a necessary consequence of the fact that no classification can be perfect. But they have accepted none of this. They have broadened the argument, counterattacked, and, finally, expelled the study of ancestors (even as fossils) not only from cladism, but from all science. That is why the subject has become controversial. We shall come to the controversial arguments; but I wish to take some other questions first. Can ancestors in fact technically be recognized? If so, how? And, if they can, how might they be classified?

The simplest cladistic technique indeed does not distinguish ancestors from sister species. Let us observe a group with the eyes of a cladist. Looking only for shared derived characters, how could we know an ancestor when we had one in front of us? Ancestors share all the same derived character states as their descendants, unless the descendant has evolved a uniquely derived character of its own (Figure 9.2a); but the same is true of sister species (Figure 9.2b). The technique confounds them; and when they are classified, both cladograms are the same (Figure 9.2c). You cannot tell from the cladogram alone whether 1 is the ancestor of 2, 2 the ancestor of 1, or 1 and 2 are phylogenetic sister groups. This only establishes what I have already asserted. Cladism, narrowly understood, does not distinguish sister species from ancestors. It also implies that the

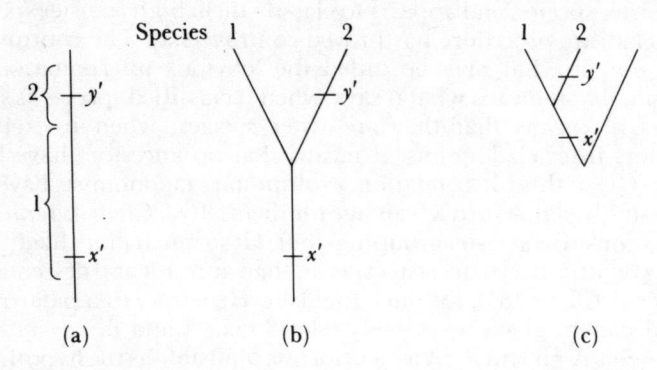

Figure 9.2
Cladograms compound sister group and ancestor-descendant relations. (a) Species 1 is the ancestor of 2. A character (x) changed state in species 1, and is shared in both, another character (y) changed state in 2 and is unique to it. (b) Species 1 and 2 are sister species. Characters, like x, that changed site in their ancestor are shared between them; characters that change in one species (like character y in species 2) are unique. (c) The cladogram in both cases is the same.

'obvious' interpretation of cladism is in error. The apparently 'sister species' of a cladogram are not in fact phylogenetic sister species. The lines of a cladogram do represent a phylogenetic relationship, but it is a more general one than the phylogenetic sister-species relationship. It stands for both ancestor–descendant pairs, and for true sister-species pairs. Cladograms do appear to represent sister species, but the appearance is unreal; the cladogram does not mean the species are phylogenetically sister species and not ancestor and descendant. Cladistic sister groups are not phylogenetic sister groups; they compound both kinds of phylogenetic relationship. Remarks like Harper's, quoted above, are not valid criticisms of cladism properly understood.

But if ancestor–descendant pairs and sister species cannot be distinguished in a cladogram, what of other methods? Let us turn now to the evolutionary taxonomist's technique. They certainly think they can recognize ancestors. To do so, they make use of the character states that cladism ignores, ancestral character states. The key fact is that ancestors must have possessed the ancestral character states for the group of their descendants. From this follows both the kind of evidence that can count for a hypothesis of ancestry, and against one. Both kinds of evidence are non-cladistic. They are, respectively, ancestral character states and uniquely derived character states. If the cladistic analysis of the characters of a group is correct, the ancestral character states of a group must, by definition, be the characters of its ancestor: the ancestor must have had all the ancestral character states; all the later derived states must have evolved from their respective ancestral states. (Zangerl (1948) called the abstraction of ancestral character states, which suggests the form of the ancestor, a morphotype.) The only other possibility is that the cladistic analysis is wrong. If a species is discovered that possesses all the ancestral states of a group, it is, so far as the evidence takes us, the ancestral one. But the evidence is only suggestive: the species is not definitely the ancestor. It may be that, for some other character that has not been investigated, the species possesses a non-ancestral state. We can be reasonably confident that we know the states of characters that we have studied, but for unstudied characters we cannot be so confident. For this reason it is almost impossible to confirm that a species is an ancestor.

The fossil record can provide further suggestive evidence. The real ancestor of a group must have preceded its descendants in time; if therefore a hypothetical ancestor is found before its hypothetical descendants, another theoretical requirement has been met, and the hypothesis corroborated. The evidence is again only suggestive. The earlier species in the fossil record was not necessarily the earlier one: the fossil record is incomplete. The two kinds of evidence are only suggestive, but when taken together they are better than nothing. We

Evolution and Classification

can advance some way to identifying a species as an ancestor. If it has all the correct ancestral characters and appears at the correct relative time, then we can say at least that it is *like* the ancestor, and exactly fits all our present knowledge of what it was like. When evolutionary taxonomists call a species an ancestor, that is all they mean, and it is pedantic to object that they do not know for sure that it is *the* ancestor. In some cases, indeed, the fossil record is so good that it is known with confidence which populations are ancestral to which (Gingerich 1979; Prothero and Lazerus 1980). But the fossil record is not usually that good and, once an apparently ancestral species has been found, a difficulty of confirmation will remain. This difficulty, I suspect, explains why many biologists feel that ancestral hypotheses commit some impropriety of scientific method.

If confirmation is here scientifically difficult, with falsification it is exactly the opposite. It is easy to prove, with an acceptable degree of confidence, that a species is not the ancestor of a group. Remember, all the characters of an ancestor must possess ancestral states. Thus, if a species possesses so much as a single derived character, it is not the ancestor. In practice we cannot place absolute confidence in our assessments of character states, and we might want three or four derived characters in a species to persuade us it was not ancestral; but that is only a practical detail. Derived character states, if they are correctly identified, provide strong evidence that a species is not an ancestor. In practice, only uniquely derived characters will be important, rather than shared derived characters. If a species shares a derived character with another species of a group, no one would ever dream that it was the ancestor of the group. They would be classified together within the group. (See species 2 and 3 in Figure 9.3.) In fact, the possession of the derived character suggests that the species is not

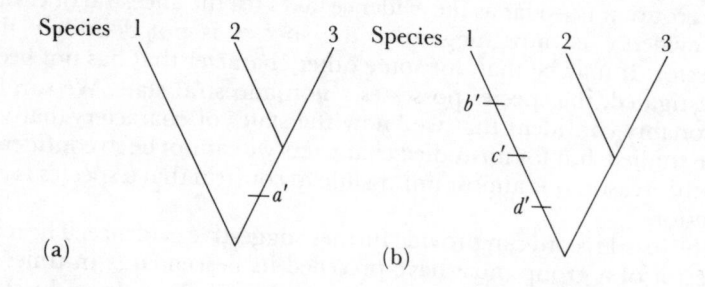

Figure 9.3
Confirmation and falsification of ancestral hypotheses. Is species 1 ancestral to species 2 and 3? Characters b, c, and d are in the ancestral state in species 2 and 3. (a) Species 1 has the ancestral states of characters a, b, c, and d, and the ancestral hypothesis is confirmed. (b) Species 1 has derived states for b, c, and d; it is falsified.

the ancestor, but that is almost too obvious to need stating. The real question concerns such species as no. 1 in Figure 9.3. It shares no derived character states with species 2 and 3: but is it their ancestor? The answer lies in those crucial characters for which species 2 and 3 have ancestral states. If species 1 is ancestral for these too (Figure 9.3a), then it is at least like the ancestor. If it has any of them in the derived state, then it is definitely not the ancestor (Figure 9.3b).

Such are the techniques by which we may test whether a species is an ancestor. In a narrow sense, they are not cladistic: neither uses shared derived character states. They do, however, draw on the same common fund of knowledge as does a cladistic classification: the states, ancestral or derived, of every character. When it is converted into a cladogram, all the knowledge of uniquely derived character states, and of ancestral character states, is, having once been discovered, thrown away. Thus in a cladogram, a species like 1 in Figure 9.3 is classified in the same place whether it has a hundred uniquely derived character states (and cannot be ancestral), or is ancestral for all of them (and might be). This information, which cladism ignores, is what is needed to distinguish ancestors from sister groups.

In summary, ancestors can be tentatively recognized by their possession of ancestral characters; derived characters prove that a species is not ancestral. Evidence therefore exists by which we can determine whether a species is not, or might be, an ancestor. The relevant evidence, however, is excluded from cladistic classifications. Given a cladogram, it is impossible to say which pairs of species are ancestor and descendant, and which share some other common ancestor.

We can fix this discussion by example. We need a case of a set of groups, of which one is thought to be the ancestor of another, and which has been studied cladistically. Several such cases have appeared recently among the vertebrates. The cladistic school has even acquired a certain notoriety as it has reorganized one major vertebrate group after another. These are not, as some would believe (Cox 1982; Panchen 1982), the noisy terrorism of cladistical public relations, but the automatic result of the system. In evolutionary taxonomy, a candidate ancestor is placed close to its descendants; other groups, with their own uniquely derived character states, are put further away. But if ancestral similarities and uniquely derived states are excluded, the candidate ancestors may become much less closely related to their descendants. Such is the method by which Rosen, Forey, Gardiner, and Patterson (1981) made the lungfish (Dipnoi), rather than the Rhipidistia, the sister group of the Tetrapoda, and Kemp (1983) made the Tritylodontidae, rather than *Probainognathus*, the sister group of the mammals. With Gardiner's (1982) reclassification of tetrapods, however, the reason is different. Here it is more his (highly questionable) character analysis than his cladistic method (see also Gardiner 1983 on the Amphibia). Either Rosen *et al.* or Kemp would illustrate the argument we have worked through. In both cases the group that had been supposed to be an ancestor has been further removed from their supposed descendants than another group that had been regarded as non-ancestral. We need look at only one to see how the candidate ancestors are dealt with by cladism. The work of Rosen *et al.* is now surrounded by too much controversy to look at here (Halstead, White, and MacIntyre 1979; Jarvik 1981): that leaves Kemp.

Kemp is concerned with the relations of the mammal-like reptiles to the mammals. We must contrast his new classification with its predecessors, which immediately raises a difficulty. There is always disagreement over classifications; and it is misleading to call any one of them orthodox. That said, let us pick on one, in the understanding that not all experts would agree with it. The work of Crompton and Jenkins (1979) will suffice. Crompton and Jenkins uncontroversially derive the mammals from the cynodonts (see Figure 9.4); but that is a large group. Within the cynodonts, the two main candidates for the ancestry of the mammals are the Tritylodontidae and *Probainognathus*. Crompton and Jenkins place the tritylodontids to one side, and *Probainognathus* nearer to the ancestor. Here is why (1979, p. 67): 'The question remains as to which group of cynodonts gave rise to the mammals. There is no clear answer at present. Some groups (e.g., tritylodontids) that acquired decidedly mammalian structures both in cranial and postcranial anatomy are, in other features (such as dentition), so specialized as to remove them from consideration.' The mammals could not have arisen from the tritylodontids because that group has uniquely derived dental conditions. The ancestor of the mammals, of course, must have possessed the ancestral dental condition. They continue: 'The Chiniquodontidae include the most advanced forms of cynodonts that did not evolve the peculiar specializations typical of tritylodontids and ictidosaurs. For the present, the chiniquodontid *Probainognathus* appears to be the closest known representative of the pre-mammalian lineage. It is unlikely, however, that *Probainognathus* was directly ancestral to the Morganucodontidae, because of the relative reduction of cingular cusps on the postcanine teeth.' *Probainognathus* is the best candidate ancestor because it contains the largest number of ancestral mammalian characters.

Such is the reasoning of Crompton and Jenkins: Kemp has examined the same characters and groups with the methods of cladism. The characters that Crompton and Jenkins used to argue a relation between *Probainognathus* and the mammals are all (according to Kemp) ancestral in the cynodonts; he duly discards them all, and finds no cladistic evidence of a relationship. But the

Ancestral groups and sister groups

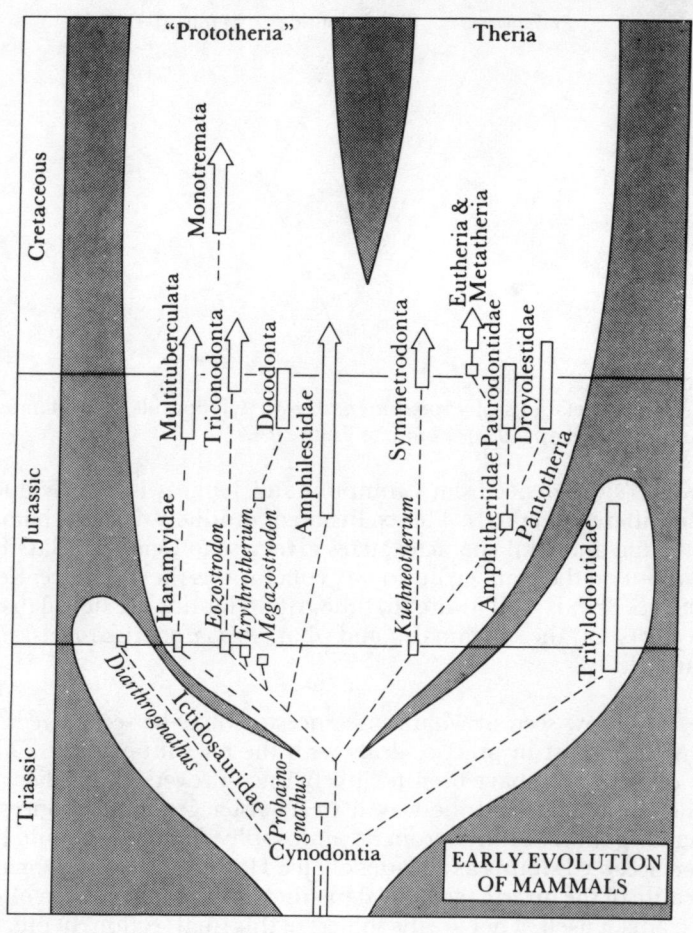

Figure 9.4
Relations of the mammals and mammal-like reptiles. (From Crompton, A. W. and Jenkins, F. A. (1979), in *Mesozoic Mammals*, eds Lillegraven, J. A. Kielan-Jaworowska, Z., and Clemens, W. A., by permission of The University of California Press.)

tritylodontids (he says) do share such derived characters with the Morganucodontidae as the condition of the periotic bone. As he discards the shared ancestral similarities of chiniquodontids and mammals, he destroys their relationship; and as he discards the uniquely derived characters of tritylodontids, he creates a relationship between them and the mammals. The group that cannot be ancestral to the mammals is thus moved closer to them than a group that could be (Figure 9.5).

Evolution and Classification

Probainognathus   Tritylodontids   Mammals

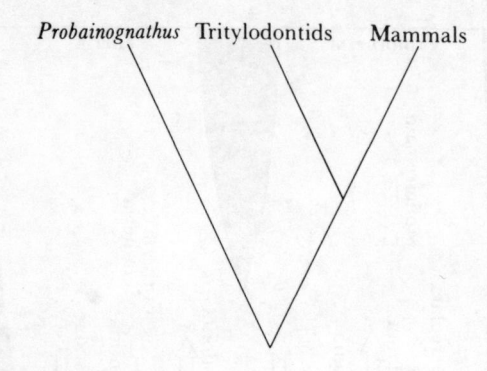

Figure 9.5
Sister-group relations of *Probainognathus*, tritylodontids, and mammals, according to Kemp (1983). Compare Figure 9.4.

Kemp's difference from Crompton and Jenkins is mainly due to their different methods. The exclusion of uniquely derived character states and ancestral character states gives rise to Kemp's reclassification. But on the main evolutionary conclusions they all agree. Kemp (1983, p. 381) is well aware that the tritylodontids are not plausible ancestors of the mammals, and that *Probainognathus* is more plausible.

We have now seen in what sense ancestors may be recognized, and how the cladist in practice deals with the relevant evidence. Their procedure could have been relatively uncontroversial. Cladism is a coherent system, and the way it treats sister groups and ancestors unambiguously follows from its philosophy. Its method could have been accepted simply as a matter of fact. However, it was not, mainly because of the bizarre theoretical justifications that have been offered by cladism itself. They are the subject of this final section. In practice, as we have seen, cladism simply ignores the relevant facts; but that is not how it sees the matter. It maintains instead that there is no evidence by which an ancestor may be recognized. As Ronald Knox would have said, cladism, having first turned its back on the relevant facts, has then had the effrontery to declare that the facts do not exist.

The case is a compound of practical and philosophical criticism. It runs something like this. Ancestral, not derived, character states characterize ancestors. As we saw in Chapter 5, there is a sense in which ancestral can be and (by certain cladists) are called 'absences'; and these 'absences' have been (wrongly) said not to exist. The conclusion is that no positive evidence can characterize an ancestor: ancestors are 'incapable of being described' (Forey 1982, p. 147). The cladistic literature insists on this point with a repetitive uniformity; but we shall let one quotation (from Farris 1976, p. 272) stand for the

146

rest. Suppose that a species S has all the ancestral character states of a group of species, A. 'It is from these observations possible that S is indeed the stem species of A. But it is also possible that ... S is in fact a constituent of A and itself possesses [uniquely derived character states] not yet observed. The sister group relationships that are properly expressed in the phylogenetic system are based only upon positive evidence. Ancestor-descendant relationships are susceptible to no such positive evidence, and should not be included.' (Similarly, Nelson 1972, p. 368; 1973, p. 311; Cracraft 1974b, pp. 73, 77; Platnick 1977, p. 440, who calls 'ignorance' what Farris calls negative evidence; and almost all the references of page 96 above; Nelson and Platnick 1984; see also Fortey and Jefferies 1982.) The reasoning is along the right lines; but the conclusion is, I think, exaggerated. The argument of Farris differs little from what we have already been through: the identification of ancestors is tentative because future study may reveal derived characters in the hypothetical ancestor. But I should draw a milder conclusion. We never know what the future may bring. If future evidence may topple our hypothetical ancestors, it may also topple our sister groups, and overturn our entire cladogram. If any hypothesis that is vulnerable to future evidence 'should not be included' then, I fear, cladism will never get off the ground. Indeed, on these terms, no science could. In fact, of course, we have to accept what our present evidence suggests, and be prepared to modify our ideas when new evidence comes in. Nor should we be deceived by that specious distinction between positive and negative evidence. It is true only of cladism, not of nature. Ancestral characters may be absent from the cladistic system, but they are not absent from animals. They exist, and they can be recognized: it is in fact a stage in the cladistic method to recognize them. Once recognized, they provide just as positive evidence of ancestralness as do the states recognized as derived, of derivation. In the end, when all exaggeration is set aside, we need not disagree much with the cladistic objection. Ancestors can be recognized, but the recognition is uncertain. It is tentative because it really is difficult to verify that a particular species is an ancestor; it is only easy to verify that one is not, but negative knowledge of that kind is not much use to the scientist. As hypotheses, ancestors are scientifically imperfect. Whether they are too imperfect, or worthy of study despite their imperfections (and remember, no hypothesis is perfect), is a matter of taste we need not settle here.

However that may be, it is worth noticing that the scientific imperfection of ancestral hypotheses is in despite of their perfect conformity with Popper's criterion of scientific respectability, the criterion of falsifiability (see also Hull 1983). Hypotheses of ancestry are easy to falsify. Falsification is accomplished (as we have seen) by derived character states. If a species possesses derived characters, it is not ancestral, and that is that. Which said, we may be surprised, while

Evolution and Classification

reading Hull (1979, p. 430), to meet a second line of cladistic objection, consisting of 'cladists [who] have argued that hypotheses about ancestor-descendant relations are unscientific because they are unfalsifiable'! Hull supplies a considerable list of references, but they are (I have found) the trail of a wild-goose chase. The cladists in that list do all try to reconcile their system with the philosophy of Popper, and they strike some singular postures as they do so; but, with one exception, none of them pronounces ancestral hypotheses to be unfalsifiable. One of Hull's culprits, Patterson (1978, p. 221) even remarks that 'common ancestry relationships can be specified by [shared derived characters], and ancestor-descendant relationships cannot. I do not think that falsification or verification has anything to do with it'. But Hull's list does contain two definite cases. Engelmann and Wiley (1977, p. 1) do indeed 'conclude that [uniquely derived character states] cannot refute ancestor-descendant relationships'. That 'cannot' is remarkable because, as we have seen, they most certainly can. How then have they persuaded themselves that they cannot? It would be disproportionate to work through their argument in full, and Wiley (1978) himself has since remarked that ancestral hypotheses are falsifiable. But the crux of their case is that the state of a character is assessed relative to a cladogram. If you falsify an ancestral hypothesis, you may drag down the entire cladogram with it. Engelmann and Wiley (1977, especially pp. 7-8) express this by saying that the ancestral hypothesis is not falsifiable within the system that the hypothesis was framed. Two points may be made. One is that, if true, the argument would show not only that ancestral hypotheses were unfalsifiable, but all of cladism as well. But the argument is not true; it fails to show that ancestral hypotheses are unfalsifiable; the fact that the cladogram may be destroyed along with the ancestral hypothesis is neither here nor there. Ancestral hypotheses can be falsified in the same sense that a cladogram can be falsified: by finding evidence that does not fit. The second point is that it may not be true that the whole cladogram is falsified when an ancestral hypothesis is falsified. As we have seen (Figure 9.3), a species is classified in the same place regardless of the number of uniquely derived characters that it possesses. If one is found in a candidate ancestor, the cladogram may be unharmed while the ancestral hypothesis falls.

In conclusion, ancestors can be recognized, by the appropriate evidence of ancestral character states, but only tentatively. The evidence of derived characters states can more conclusively show that a species is not an ancestor. It happens that the evidence that is used to confirm, and to falsify, ancestral hypotheses is not included in cladistic classifications, which do not distinguish ancestor-descendant pairs among the sister groups of the cladogram. In Kemp's cladogram of mammalian relatives (Figure 9.5), for example, the further sister

Ancestral groups and sister groups

group from mammals is more likely to be the ancestor of mammals than is the nearer sister group. There is nothing wrong in this. We may wish our classification to represent everything, but it is a dangerous desire, and must be overcome. A classification cannot represent all kinds of information without becoming ambiguous. Cladism does confound sister group and ancestor–descendant relations; but the system as a whole is soundly principled. Unfortunately, the transforming system of cladism has also sought to establish that ancestors cannot be studied scientifically. But they can be: there is no reason why cladistic and ancestral hypotheses should not both be studied. They do both merit study. Although the difference between ancestral and sister-species relations cannot be represented in a cladogram, it is important.

10
Cladograms and speciation

'The view is often advanced that phylogenetic systematics presupposes a dichotomous structure of the phylogenetic tree', wrote Hennig (1966, p. 209); and he then went on to defend his preference for dichotomous classification. Most of his followers, transformed and untransformed, share that preference. Given that the preference exists, it is the purpose of this chapter to consider how it might be justified. Whether or not one should aim for dichotomous classifications may seem a minor point to argue over at such length; especially as the argument will be entirely theoretical, even though, if this chapter is correct the crucial argument would be not theoretical but factual. The important question, I shall seek to show, is what the real pattern of speciation is; but having argued as far as that I shall not go on to ask what the pattern of speciation is. I am concerned here with the point of principle. For large questions of principle may be discussed in minor issues. The large question is whether cladistic classification can be separated from the theory of evolution. The issue of dichotomous cladograms has become a test case, extensively discussed (see Hull 1979, pp. 425-7 for reference to the controversy). The cladistic procedure is not in doubt: it is to seek dichotomous classification. The controversial question is whether it can be justified without the theory of evolution. We can consider a related question at the same time, that of whether the preference is a necessary, or dispensable, part of cladism.

Cladograms, as we have seen, were originally intended to represent the phylogenetic relations of species. The obvious interpretation, here as always, of a cladogram is that it means what it says: if, for instance, it is dichotomous, it means that the pattern of branching among the represented species was dichotomous; the species arose two at a time, with the ancestor going extinct (as it must, by definition) in the act of ancestry. A preference for dichotomous cladograms should then mean that ancestral species generally give rise to two descendants at a time. In the case where an ancestor does not change its appearance after splitting, the preference should mean that ancestral species generally bud off descendants one at a time (which is cladistically equivalent to dichotomous speciation because the ancestor changes species every time it buds off a descendant). Such is the obvious interpretation: but how have the schools of cladism themselves justified the preference? The obvious interpretation might serve in Hennigian cladism; but it cannot in transformed cladism because theories about speciation are 'process' theories, which are banned from that system. They will have to find other reasons.

Hennig's own argument has two parts (1966, pp. 209–13). Suppose first that the phylogenetic relations are dichotomous, and the evidence agrees with nature, the pattern of shared derived characters suggesting a dichotomy. There is then no difficulty, for the preference for dichotomous classification is simply a preference for the truth. The difficulty arises in the second case, when the evidence does not suggest a dichotomy. If a dichotomous pattern is still to be preferred then, in the face of the facts, some other justification will be needed. Hennig proceeds in two stages. He first points out that there are two interpretations when the facts suggest a polytomy (which I shall here use to mean a branch point with more than two branches): either speciation really was polytomous; or it was really dichotomous, and the facts so far available are misleading. Polytomies may be 'unresolved dichotomies'. He then offers a reason in favour of the second interpretation. 'A priori [he says] it is very improbable that a stem species actually disintegrates into several daughter species at once.' Non-dichotomous speciation, in other words, is improbable, which is what I have called the obvious interpretation over again. Characters studied in the future are likely to fit the pattern of speciation. If speciation is dichotomous, future characters are likely to indicate a dichotomous classification, and dissolve the polytomy into a nest of dichotomies. Therefore, even when the facts do not suggest a dichotomy, a preference for dichotomous classification is still a preference for the truth if speciation generally is dichotomous. Hennig continues, however, by stressing that, at this fine level of resolution, it is all very difficult. 'Here [cladism] is up against the limits of the solubility of its problems. These limits are set by a certain vagueness and indeterminacy of the concept of "simultaneous".'

The two steps of Hennig's second argument stand and fall together. Polytomies may be unresolved dichotomies, but only if speciation generally is dichtomous. I wish to spell this point out in more detail. Hennig's quoted theory of speciation really is a crucial prop of his preference for dichotomous classification. Hennig himself did not prove it, and (I suspect) was not fully aware of the dependence; certainly many biologists are not, as the preference for dichotomies is often justified by the 'unresolved' status of polytomies, as if this were

Evolution and Classification

independent of the pattern of speciation. Polytomies may indeed be liable to dissolution by further study; but they also may not. Why should we expect one or the other? I shall argue that only if the actual pattern of speciation is dichotomous will polytomies tend to dissolve. This I shall try to prove by examining the circumstances under which, first a trichotomy, and second a dichotomy, are vulnerable under further study. The same point has been made more briefly by Hull (1979, p. 426).

Suppose, then, that we have studied some characters in a set of species, and the pattern of shared derived states suggests a trichotomy. Why might further study dissolve it into a pair of dichotomies? The obvious reason is the pattern of speciation. If it is dichotomous, shared derived characters ought to fall into paired groups, according to the dichotomous splits. We could reasonably suppose that the characters we have studied so far are unrepresentative: subsequent characters should fit the dichotomous pattern, and justify our preference. So far, so good; but now suppose that the species really did split trichotomously. Will further study in this case resolve the trichotomy into a pair of dichotomies? It will not. The future facts should resemble the known ones. As indeed they should, because the shared derived characters were after all meant to reveal the pattern of branching. The dichotomous preference is now in error, and a trichotomous preference would be appropriate. Likewise, if the actual pattern of branching was trichotomous, and the provisional cladogram was dichotomous, further study should break down the dichotomy and resolve it into the (natural) trichotomy. Let us fix this in an imaginary example. Suppose that three species were formed at a

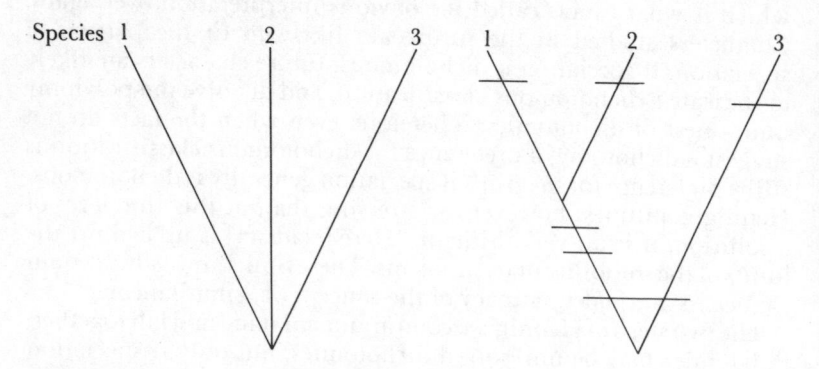

Figure 10.1
(a) The real pattern of evolution was a trichotomy of an ancestor into species 1, 2, and 3. Initial cladistic study (b) of six derived character states suggests a dichotomy. As usual, derived character states are indicated by horizontal lines: if a continuous line crosses two species, they share the same character state.

Cladograms and speciation

trifurcation, but study of the first few characters happened to suggest a pair of bifurcations (Figure 10.1). We study some more characters, which will, in all probability, falsify the dichotomies. Now the earlier testimony will be exposed, and some of what were previously thought to be shared derived characters will be revealed to be convergences, or incorrectly analysed in some other way. The dichotomies will have to be dismantled and replaced by a trichotomy (Figure 10.2). Without its second part (the assumption that speciation is dichotomous), the first part of Hennig's argument can be stood on its head. We might just as well say that we cannot distinguish a nest of dichotomies from an 'unresolved' trichotomy (or tetratomy, and so on). If we do not assume that speciation is dichotomous, there is no reason to expect further cladistic study to produce more and more dichotomous cladograms. Dichotomous cladograms are the natural end point of cladism only if speciation is dichotomous.

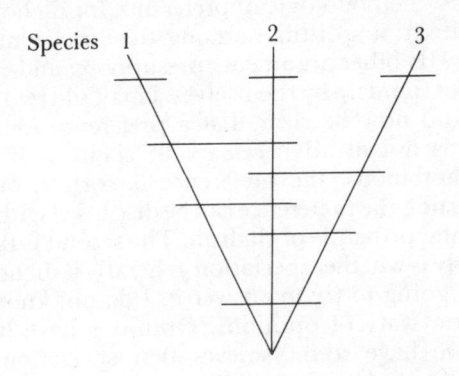

Figure 10.2
Study of another two characters (after Figure 10.1b) in species 1, 2, and 3 breaks down the dichotomous cladogram, and resolves it into a trichotomy.

Hennig makes another point. After remarking that apparent polytomies only portray our uncertain knowledge, he writes (p. 213):

> This raises a question. In younger monophyletic groups it is evidently impossible to distinguish with certainty between [polytomy] and dichotomy, or to recognize a probable consistent dichotomy as such. Should we not then question the clear dichotomy that seems to be recognizable in so many older groups (Acrania-Craniota, Cyclostomata-Gnathostomata, [etc.])? May not the picture of a [polytomy]... be the correct one for these older groups too, and the picture of a clear dichotomy actually only be an artefact of our methods

153

of recognizing kinship relations? In my opinion this need not be feared.

But the reason he then gives is a new argument completely. He now points out that even with an underlying pattern of trichotomies, random extinction will tend to reduce them to dichtotomies (Hennig 1966, figs 64 and 65). The dichotomous classification then need not fear the future, because the other species will not be found (the study is assumed to embrace only modern forms). But the argument only applies if one of the groups of a trichotomy has gone extinct; then the real pattern of the species under study has become dichotomous again, and the preference (as before) is correct. The true problem for the preference is when the real pattern of the species under study is trichotomous. Then a present dichotomy has much to fear from future study. It can expect to be resolved into a trichotomy.

In summary for Hennig, the only valid justification that he presents for his 'methodological' preference for dichotomies is that the actual pattern of splitting, among the species under study, is dichotomous. His other arguments presuppose, and reduce to, this one: they do not stand up by themselves. I may add two remarks. One is that it should now be clear that a preference for dichotomous classification is not at all necessary in cladism. If speciation is generally dichotomous, the preference is correct, and should be applied. If it is not, the preference can be discarded without violating any fundamental principle of cladism. The second is that the crucial question clearly is whether speciation generally is dichotomous. I am not, as I said, going to try to answer it. I do not know of a factual review, and am wary of opinions. Opinions have been declared. Hennig, as we have seen, believes that speciation is generally dichotomous: Simpson (1978, p. 271) on the other hand says this is 'not true'. I incline to Hennig's opinion. Given the amount of time available, and the number of species there have been, perhaps the evolution of new species is so rare an event that more than two new species would not often have evolved from one ancestor at the same time. But then there is the thought that geographical variation is a source of new species, and one species is often split into many geographical races (Mayr 1963). A proper analysis would have to make quite nice distinctions about the degrees of interbreeding, which will control the chance that a new character becomes shared among races; 'speciation' is too crude a concept at this fine level. But although the question of fact is interesting, it is not the issue here. I have sought only to show that if cladism is to prefer dichotomous cladograms, it must justify that preference by the pattern of speciation.

Let us turn now to transformed cladism. Our main source will be Platnick (1979). As a transformed cladist he is naturally worried that

Cladograms and speciation

Hennig's arguments 'reflect a concern with the mechanism of speciation' (p. 539). He would not justify cladistic procedures by biological argument, and 'the question with regard to the methodological preference for dichotomous hypotheses, therefore, [becomes] whether the preference can be justified by arguments that do not depend on any particular view of the mechanism of speciation.' He has convinced himself that it can: he has found a justification, but it is not in biology ... it is in philosophy.

Platnick justifies a preference for dichotomous cladograms by their greater supposed 'information content'. To see what this means, let us, with Platnick, compare a hypothetical trichotomy of three species with a pair of dichotomies (Figure 10.3). The pair of dichotomies, he believes, has a greater information content; it predicts more kinds of derived characters. The trichotomy (he says) predicts four kinds of derived characters: they can be unique to any one of the three species (predictions nos 1–3) and to all three (no. 4). The pair of dichotomies makes all these four predictions and a fifth: shared derived characters between the sister-species pair. 'This additional information content is enough for a preference for dichotomous hypotheses, without any recourse to any knowledge claims about the mechanism of speciation' (Platnick 1979, p. 540). Is it indeed? If his argument is to work, two things must be true: dichotomous cladograms must be more informative than (say) trichotomies; and, in any case, greater 'information content' must be a good reason to prefer a hypothesis. Let us see whether these two conditions can be satisfied.

First, then: are dichotomies more informative? Platnick has achieved his '25% greater information' in the dichotomy by excluding the negative predictions made by the trichotomy. In fact, every extra

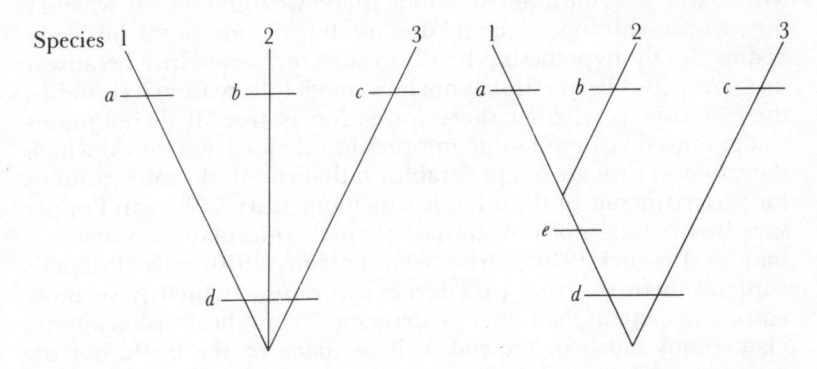

Figure 10.3
Information content of dichotomous (a) and trichotomous (b) cladograms, redrawn like Platnick (1979, fig. 1). The trichotomy, according to Platnick, predicts only four kinds ($a-d$) of shared derived characters, whereas the dichotomous cladogram predicts five ($a-e$). The present text disagrees.

positive prediction made by the pair of dichotomies is matched by a negative prediction in the trichotomy. If the dichotomy predicts shared derived characters between species 1 and 2, the trichotomy contradicts them. There may be some, but in that case any apparently derived characters (probably in fact convergences) shared between species 1 and 2 should be no more common than characters shared between species 1 and 3, or 2 and 3 (cf. Platnick 1979, p. 540). The trichotomy makes the same total number of predictions: it contains the same amount of information. (Nelson's (1979, p. 5) definition of 'information content' likewise excludes the negative predictions of a cladogram.)

Suppose, however, that we asserted that predictions have to be positive, not negative (which would be nonsense, but let that pass). Or just suppose that the previous paragraph did not exist. We can then ask whether greater information content is a reason for preferring a hypothesis. The answer is that it is not. A hypothesis is not preferable simply because it contains more information; hypotheses are preferred because they are true. A paired dichotomy with species 2 and 3 as sister groups would contain just as much transformed cladistic information as one with species 1 and 2 as sister groups (Figure 10.3), but the information is very different. By the criterion of information content, the two would be equally good hypotheses. But they should be decided between by the facts: the hypothesis should be chosen that fits them best. Obviously a hypothesis could be made which contained even more information. We could dream up an exactly specified branching pattern for all existing species, and specify, furthermore, the exact time of branching. We could pile in yet more information. We could say where, and how, the branching took place: we could specify whether the sun was shining: we could declare the position of the heavenly bodies. But the hypothesis is hardly to be recommended just because it says so much. The question is not how much information it contains, the question is whether the information is true. If dichotomous cladograms do contain more information than trichotomies (which they do not) they are not preferable on that count: the only grounds for preferring one or the other is which one is true. Nor can Popper save this criterion of transformed cladistic information content. It may be (Platnick 1979, p. 540) 'what Lakatos (1970) calls "Popper's supreme heuristic rule [to] devise conjectures which have more empirical content than their predeceesors"', but the decision among cladograms must, in the end, still be made by the truth, not the quantity, of the 'empirical content'.

The transformed cladistic argument thus fails. The preference for dichotomous cladograms has not been justified on any other grounds than that speciation is dichotomous. Of course, speciation may indeed be dichotomous. It is not my case that speciation is not

Cladograms and speciation

dichotomous, nor that dichotomous cladograms should not be preferred. I am not criticizing the preference for dichotomous cladograms, I am criticizing the transformed cladistic justification of it. My point is that a theory of speciation – an evolutionary theory – is needed to justify the preference.

At an earlier stage, the cladistic preference for dichotomous classification was justified by the pattern of speciation, and many critics duly criticized it by arguing that the pattern may often not be dichotomous (Mayr 1974; Simpson 1978; references of Hull 1979). Cladists reply (Nelson and Platnick 1980) that this misses the point: they do not justify the preference by a theory of speciation, but by philosophy. Perhaps some critics have missed the point of transformed cladism. But then perhaps they were aware of the philosophy of transformed cladism, and silently ignored it out of ... well, whatever, or because they knew the only hope of a justification lay in the pattern of speciation. The purpose of the present chapter has been to prove, for one more cladistic principle, that philosophy does not support the transformed cladist. The justification must be found in the process of evolution.

11
Evolution and classification

What is the correct relation of classification and evolution, of (in the terms of transformed cladism) 'pattern' and 'process'? The central doctrine of transformed cladism (as we have seen) is that the classification of organic diversity must be kept separate from the theories that may explain it, which means in particular that the theory of evolution must be kept out of the procedures of classification. Transformed cladists have expressed the point briefly by saying pattern and process must be studied separately, pattern before process. Another quotation, from Cracraft (1983, p. 164) this time, will re-explain the terms: 'it is becoming increasingly apparent that systematics is the area of biology that defines the *pattern* of organic change through space and time and, consequently, specifies that body of knowledge that theories of evolutionary *process* must be capable of explaining.' So confident are many transformed cladists of their distinction of pattern and process that they even scold their critics with it. Thus, the 'so many...commentators who criticize transformed cladistics' learn from Humphries (1983, p. 305) that 'they confuse the cladistic activity of ordering characters to their level of universality with a causal explanation of the hierarchy'. If it were an error to suppose the two are connected, this would be a good reply. I do believe classification and evolution are strongly connected; my aim has not been to confuse them unwittingly but to fuse them together with argument. We have examined the doctrine of transformed cladism in its many applications, and seen how evolution and classification are connected in particular places: I now wish to conclude by stating the argument both in its abstract form, and in the terms of transformed cladism itself.

The double distinction of pattern and process has a closely related triple form, the distinction of 'cladograms', 'trees', and 'scenarios' (Eldredge and Tattersall 1977; Cracraft 1979; Eldredge 1979; Eldredge and Cracraft 1980). A cladogram is a branching diagram of species, clustered by their shared derived characters; the lines among the species indicate only the sharing of derived characters. A tree is a phylogeny, with ancestral species placed at the branching nodes, rather than at the tip of a branch, as in a cladogram; the lines among species represent evolutionary lineages. A 'scenario' is an account of the evolutionary process that causes the tree and its cladogram. The distinction is real, but it has been put to a transformed cladistic use which we must oppose - the doctrine that, as a matter of method, the cladogram must be constructed first, and only then can one go on (if one so chooses) to construct a tree, and even try to explain it. The flow of information would then be strictly one-way: a cladogram would be needed before a tree could be suggested, a tree before a 'scenario' (references above, and those cited in Hull 1980, p. 132; see also Fisher 1981 and Platnick, quoted on p. 18). Because the cladogram is the 'pattern', and the scenario the 'process', in the triple distinction, as in the double, the theory of evolution is separated from classification. Because the cladogram-tree-scenario distinction is so closely related to the pattern-process one, we can take them together. In both, transformed cladism proposes to expel the theory of evolution from cladistic classification.

Previous chapters have accumulated four sources of opposition. The first, and one of overwhelming importance, is that evolution is needed to justify the whole system (Ch. 2-7). The remaining three matter less, and are, secondly, that natural selection can provide a cladistic technique (Ch. 8), and thirdly, that both trees and cladograms are built from the same common fund of knowledge (Ch. 9); the fourth argument matters least, because the philosophy of cladism does not demand dichotomous classifications, but it is that, if a dichotomous preference is to be maintained, it must be justified by a theory of speciation (Ch. 10). A fifth criticism may be taken from Hull (1979, 1980), and added to the list: even if the distinctions of pattern from process, and of cladogram, tree, and scenario, were true in philosophy, they would not necessarily, for that reason alone, be true in practice. Let us examine these points, taking Hull's first.

Let us pretend, for sake of argument, that the distinctions are philosophically valid. Let us suppose that classification is epistemologically separate from, and prior to, evolutionary explanation. Transformed cladism would then deduce the practical conclusion that we must study pattern before process, and build cladograms before trees, and trees before scenarios. Hull (1980, p. 132), however, remarks 'of course, no taxonomist has ever proceeded in this fashion, and according to the philosophy of science explicitly espoused by the cladists, no necessary or even important connection exists between either logical or epistemological priority and the actual temporal order in which scientists perform their task (Gaffney [1979], pp. 79-84)'. In other words, even were classificatory pattern philosophically prior to evolutionary process, the practical conclusion would not necessarily follow. But is the philosophy itself valid? How, in practice, should the theory of evolution be used in classification? *Naturam expelles furca, tamen usque recurrat.* The branching hierarchy of phylogeny, as Hennig realized, is the only unique, unambiguous hierarchy in nature that can be the aim of classification. It is the only known objective system. When phenetic classification discards evolution, it becomes subjective; when 'evolutionary' classification overrules the phylogenetic principle, it becomes subjective; and when transformed cladism discards evolution, it too becomes subjective. The way forward for all three schools, as each is well aware, is towards objective classification, towards unambiguous natural hierarchies. That means embracing, embracing more keenly, or re-embracing, the theory of evolution and the branching hierarchy it implies. They must become cladistic, in the evolutionary, untransformed and reformed, sense. Evolution, in Hennig's argument, is needed to justify cladism. Even if that theory of process can be kept out of the methods, as they are actually operated to detect patterns, it cannot be kept out of any justification of those methods. Cladists may not think about evolution while they apply such methods as the embryological criterion, or outgroup comparison; but they need to as soon as they try to explain why they are using those techniques. The theory of evolution has a further use. It is needed to improve, or to modify, the cladistic techniques. Whenever the application of a cladistic technique to a character is not perfectly straightforward, the cladist should return to the principle from which the method was developed. It is from there that the technical improvements must come, and only there that they can be justified. For example, an understanding of the circumstances that favour neoteny (p. 67) would help to sort out the paradoxes of the embryological criterion.

Philosophical justification, and technical improvement, thus flow from the theory of evolution into classification, menacing the transformed cladist's distinction on the way. The transformed cladist might still plead that the methods, at least in their most mundane, operational work do not contain a theory of process. But this too would be a mistake. All the cladistic methods stem from the theory of evolution. Outgroup comparison makes an assumption about the frequency of convergence; the embryological criterion assumes that new evolutionary stages are always added on at the end of the adult stage; and there would be little reason to suppose that ancestral stages appear earlier in the fossil record if they did not evolve before derived character states. With the functional criterion, the dependence on a theory of process, natural selection in this case, is particularly clear. Evolutionary theory is put to particularly active use in this method. Each character has to be referred back to the theory of natural selection, which acts as a continual reminder of how classification uses the theory of evolution. Information here will obviously be flowing from the theory of evolution into the classification: from 'scenario' to 'cladogram'. The 'scenario' will have to be understood in order to construct the cladogram (Fisher 1981). Cracraft (1981, p. 22) believed that the functional criterion 'confounded process analysis with pattern recognition', and was therefore in error. His premise is correct, but the conclusion upside-down. The functional criterion does indeed put together pattern and process, but it is the pattern-process distinction, not the functional criterion, which comes out the worse.

If the functional criterion so clearly damages the cladistic distinction of cladograms and scenarios, what about the middle part of the triple distinction? What about trees? A tree differs from a cladogram in that ancestors are tentatively recognized. A tree is consistent with several cladograms – with several hundred in a typical case (because ancestors may appear in many places in a cladogram, but in only one place in a tree) – and is in that sense a more exact theory. It might therefore be thought that, to convert a cladogram into a tree, we should need extra knowledge, and that cladograms must be discovered before trees. Transformed cladism argues exactly that. I would not, however. The 'extra' knowledge mainly concerns which character states are ancestral, because we need to know whether particular species in the cladogram possess ancestral or uniquely derived character states. If they have ancestral states, they may be ancestors; if derived, they cannot be. That is the 'extra knowledge'. It is not 'extra' at all. It is obtained during the initial cladistic analysis of character states. In discovering which character states are derived, cladism has to discover which are ancestral. It does then throw that knowledge away before constructing its classification, but that is its own business. Cladograms and trees draw on the same common pool of knowledge, and could be built at the same time. Actually, whether characters are ancestral or derived is only the most important knowledge needed to convert a cladogram into a tree. Palaeontology can also be useful. A species is more likely to be an ancestor if it preceded its hypothetical descendants in time. But that knowledge comes out of the cladistic analysis too! The palaeontological criterion for distinguishing ancestral from derived character states will have discovered all the relevant facts about temporal priority. Cladogram and tree can be built at the same time, from the same knowledge. I therefore conclude that the cladistic distinction of pattern from process (or of cladogram from tree, from scenario) is in error. Cladograms have neither technical, nor temporal priority to trees. They can be built at the same time, from the same cladistic knowledge. 'Scenarios' sometimes may be needed before cladograms and trees, and sometimes they may not. (Hendy and Penny 1984 offer another reason against distinguishing cladograms from trees.)

Notwithstanding all that, the triple distinction of cladogram, tree, and scenario does contain some truth. Its legitimate meaning is that as we move from cladogram, to tree, to scenario, we move to hypotheses of increasing precision. A cladogram is consistent with several trees, and a tree is consistent with several scenarios. In this sense, 'more knowledge' is needed at each successive step. In the case of the step from cladogram to tree, that 'more knowledge' is disingenuously so called because the knowledge was acquired in the cladistic analysis, only to be thrown away before building the cladogram. No such general remark can be made for the next step, to the 'scenario'. The form of the distinction in terms of the greater precision of the theories at each stage is valid; but it does not justify either the philosophy, or practical conclusion, of transformed cladism.

We have often observed how transformed cladism finds the theory of evolution unnecessary. Unnecessary, however, does not necessarily mean undesirable in all cases, and Platnick (1979, pp. 544–5) finishes his paper by asking what the proper relation of evolution and classification might be in transformed cladism. Here is his answer: 'presumably, its role is exactly as early evolutionists perceived it, namely, as an explanation for the existence of a natural hierarchical system'. For Platnick, this evolutionary rationale is just a surface gloss which the taxonomist, after his work is done, may affect, take up or put down, according to his wish; it is an 'unnecessary' assumption. There is another similarity with numerical taxonomists here; they too used to declare that phenetic relations must be worked out first, and then, when that is done, the groups may be interpreted, by evolution if you wish, as a later affectation (Hull 1979, p. 423n, gives references).

But for the Hennigian cladist, evolution is much more crucial than that. Cladism needs evolution in its techniques, although not actually to operate them in the day-to-day practical sense. It needs evolution in its philosophy. The theory of evolution is necessary – not just desirable or decorative – in the justification of cladism. The theories of evolution and of natural selection are part of the technical machinery of the system, for they are the source of new methods, and the court of appeal for the old; but the main contribution of the theory of evolution is to underwrite the philosophy of the whole system. The investigation of classificatory patterns without the theory of evolution is doomed to be, in technique stagnant and impoverished, in practice hypocritical, in philosophy incoherent and subjective.

12
Conclusions

1. The fact that different characters of organisms define different groups has forced taxonomists to develop, and justify, principles of classification. Otherwise classification will be subjective. Taxonomists have to choose what kinds of groups they desire in their classifications: they have to choose which kind of characters to use. Two main principles, and from them three main schools, of classification have been developed: the phylogenetic and the phenetic principles; and the evolutionary, phenetic, and cladistic schools of classification. Each seeks natural, rather than artificial, and objective, rather than subjective, classification.

2. Each school contains a set of practical techniques, to choose characters and to convert shared characters into a classification, together with a philosophical justification of why their techniques should be used rather than those of the other schools.

3. The justification of evolutionary taxonomy is that evolution is the cause of the natural system of classification; natural classification therefore will represent the hierarchy of evolution. Its techniques are designed to discover the evolutionary hierarchy. They have to distinguish homologies (characters of common ancestry) from analogies (non-ancestral similarities). The techniques are practical, but imperfect.

4. The evolutionary hierarchy has two aspects, order of splitting and phenetic divergence. Evolutionary classifications represent both of them. They can, however, conflict, when divergent phenetic change is exceptionally rapid, and when the phenetic change is convergent. Evolutionary taxonomists then have to choose whether to represent the phenetic or the phylogenetic relationship. They choose phenetic similarity when phenetic

163

Evolution and Classification

change is divergent, and phylogenetic similarity when it is convergent. This choice has not been properly justified.

5. The justification of numerical phenetic taxonomy is its repeatability and objectivity. It would eliminate subjectivity from classification by choosing a large number of characters at random, and erecting its classifications by means of a hierarchic multivariate statistic. The classification represents aggregate (or 'overall') phenotypic similarity.

6. The aggregate phenotypic similarity among a pair of species depends on the statistic used to measure it: it has no objective, natural existence. There are many measures of it, and they give different classifications. The phenetic taxonomist has to choose among them. This choice is subjective. Although the procedure, once a statistic has been chosen, is repeatable, the choice itself is subjective. The whole system is subjective. The philosophy of numerical phenetic taxonomy is unsatisfactory. Because evolutionary taxonomy sometimes classifies phenetically, it too partly suffers from the same defect.

7. The justification of cladism is that the branching hierarchy of phylogeny is the only natural hierarchy suited to classification. It is unique and therefore unambiguous. Cladism technically discovers the order of branching of species by shared derived character states. It possesses techniques to distinguish ancestral from derived character states. Outgroup comparison, von Baer's embryological law, and palaeontology are three. All of them are practical and useful, none is perfect.

8. Cladism, which possesses both a sound philosophy and a set of practical techniques, is the best justified of the three schools. Its coherence is guaranteed by its philosophy, and its philosophy assumes that evolution is true.

9. Transformed cladism, a new and fourth school, would operate the cladistic techniques but deny their phylogenetic justification.

10. Four possible reasons have been offered for why cladistic techniques should be used without the assumption of evolution. One is that it can be done; which is true but not a justification, for the cladistic techniques are only one set among an infinity of practical classificatory techniques. It would leave the school subjective. The second proposed justification suggests that only the cladistic techniques can define groups at all. It is a verbal trick.

Conclusions

11. The third proposed justification of transformed cladism is that taxonomists before Darwin classified without evolution, which shows that it must still be possible. It is indeed possible, but that is no justification. No pre-Darwinian system of classification used a principle that is still valid today: those systems therefore cannot justify transformed cladism. Nor did taxonomists before Darwin use cladistic techniques. Many used techniques more like those of evolutionary taxonomy.

12. The fourth proposed justification of transformed cladism is that evolution must not be assumed in classification in order that the classification itself may be used to test the truth of evolution. Classifications can test the theory of evolution, but that does not justify transformed cladism, because: (1) classification is not needed to test evolution; (2) the techniques of all three schools, not just of cladism, could test evolution; and (3) to assume evolution in classification and to use classification to test evolution is not a viciously circular argument.

13. Transformed cladism should be rejected. It possesses practical techniques but no reason why they in particular should be used.

14. In addition to the extreme transformation of cladism to form a whole new school, there are partial transformations of cladism which remove the theory of evolution from places where it is needed.

15. The theory of natural selection can, in the form of a 'functional criterion', provide a cladistic technique to distinguish homologies from analogies, and to distinguish ancestral from derived character states, and therefore to reconcile contradictory characters.

16. Hypotheses concerning whether a species is ancestral to another species (or group of species) can be tentatively confirmed by ancestral character states, and decisively rejected by derived character states. Ancestral hypotheses can be scientifically studied. The relations among species in a cladogram compound the sister-species relationship with the ancestor–descendant one. This is not a strong objection to cladism.

17. Cladism need not prefer its classifications (cladograms) to be dichotomous. But if dichotomous cladograms are to be preferred, when the evidence does not suggest them, the preference must be justified by a theory of speciation. It cannot be justified by the relative 'information content' of dichotomous cladograms and trichotomous cladograms, which are equivalent.

Evolution and Classification

18. The best known system of classification, Hennigian cladism, uses evolution as an assumption to justify its techniques of character choice. Evolution and natural selection are its source of ideas. The investigation of classificatory 'patterns' should employ theories of the evolutionary 'process'.

References

ADLER, J. and J. CARY (1982). Enigmas of evolution. *Newsweek* **99** (29 March), 40-4.
AGASSIZ, L. (1859). *An Essay on Classification.* Longman, Brown, Green, &c., London.
ASHLOCK, P. H. (1971). Monophyly and associated terms. *Systematic Zoology* **20**, 63-9.
ASHLOCK, P. H. (1972). Monophyly again. *Systematic Zoology* **21**, 430-7.
ASHLOCK, P. H. (1980). An evolutionary systematist's view of classification. *Systematic Zoology* **28**, 441-50.
AYALA, F. J., M. L. TRACY, L. G. BARR, J. F. McDONALD, and S. PEREZ-SALAS (1974). Genetic variation in natural populations of five *Drosophila* species and the hypothesis of the selective neutrality of protein polymorphisms. *Genetics* **77**, 343-84.
BAER, K. E. von (1828). *Ueber Entwicklungs-Geschicht der Tiere,* 5th Sch. Bornträger, Königsberg. (Translated 1853, Fragments relating to philosophical zoology, *Scientific Memoirs (Natural History),* pp. 176-238.)
BALME, D. M. (1961). Aristotle's use of differentiae in zoology. In S. Mansion (ed.), *Aristote et les problèmes de méthode,* Louvain, pp. 195-212. (Revised version 1975, in J. Barnes, M. Schofield, and R. Sorabji (eds), *Articles on Aristotle 1. Science.* Duckworth, London, pp. 183-93. Page references in the text are to the revised edition.)
BALME, D. M. (1972). *Aristotle's De Partibus Animalium I and De Generatione Animalium I.* Clarendon Press, Oxford.
BEATTY, J. (1982). Classes and cladists. *Systematic Zoology* **31**, 25-34.
BENSON, W. W. (1972). Natural selection for Müllerian mimicry in *Heliconius erato* in Costa Rica. *Science* **176**, 936-9.
BIGELOW, R. S. (1956). Monophyletic classification and evolution. *Systematic Zoology* **5**, 145-6.
BIGELOW, R. S. (1958). Classification and phylogeny. *Systematic Zoology* **7**, 49-59.

BLACKWELDER, R. E. (1962). Animal taxonomy and the new systematics. *Surveys of Biological Progress* 4, 1-57.

BOCK, W. J. (1963). Relationships of the ratites based upon their skull morphology. *Proceedings of the XIII International Ornithology Congress*, Ithaca, New York, pp. 120-55.

BOCK, W. J. (1965). The role of adaptive mechanisms in the origin of higher levels of organization. *Systematic Zoology* 14, 272-87.

BOCK, W. J. (1974 [1973]). Philosophical foundations of classical evolutionary classification. *Systematic Zoology* 22, 375-92.

BOCK, W. J. (1981). Functional-adaptive analysis in evolutionary classification. *American Zoologist* 21, 5-20.

BOWLER, P. J. (1977). Darwin and the arguments from design: suggestions for a reevaluation. *Journal of the History of Biology* 10, 29-43.

BOWLER, P. J. (1983). *The Eclipse of Darwinism.* Johns Hopkins University Press, Baltimore.

BRADY, R. H. (1982). Theoretical issues and 'pattern cladistics'. *Systematic Zoology* 31, 286-91.

BRODA, P. (1979). *Plasmids.* W. H. Freeman, San Francisco.

BROWER, L., L. M. COOK, and H. J. CROZE (1967). Predator responses to artificial Batesian mimics released in a neotropical environment. *Evolution* 21, 11-23.

BRUNDIN, L. (1972). Phylogenetics and biogeography. *Systematic Zoology* 21, 69-79.

CAIN, A. J. (1958). Logic and memory in Linnaeus's system of taxonomy. *Proceedings of the Linnean Society* 169, 144-63.

CAIN, A. J. (1959a). Deductive and inductive methods in post-Linnean taxonomy. *Proceedings of the Linnean Society* 170, 185-217.

CAIN, A. J. ed. (1959b). *Function and Taxonomic Importance.* Publ. no. 3, Systematics Association, London.

CAIN, A. J. (1962). The evolution of taxonomic principles. *Symposia of the Society for General Microbiology* 12, 1-13.

CAIN, A. J. (1964). The perfection of animals. *Viewpoints in Biology* 3, 36-63.

CAIN, A. J. (1967). One phylogenetic system. *Nature, London* 216, 412.

CAIN, A. J. (1982). On homology and convergence. In K. A. Joysey and A. E. Friday (eds), *Problems of Phylogenetic Reconstruction.* Academic Press, London, pp. 1-19.

CAIN, A. J. and G. A. HARRISON (1958). An analysis of the taxonomist's judgement of affinity. *Proceedings of the Zoological Society of London* 131, 85-98.

CAIN, A. J. and G. A. HARRISON (1960). Phyletic weighting. *Proceedings of the Zoological Society of London* 135, 1-31.

CARSON, H. L. (1983). Chromosomal sequences and interisland colonizations in Hawaiian *Drosophila*. *Genetics* 103, 465-82.

References

CARSON, H. L. and K. Y. KANESHIRO (1976). *Drosophila* of Hawaii: systematics and ecological genetics. *Annual Review of Ecology and Systematics* **7**, 311-45.

CARSON, H. L. and J. S. YOON (1982). Genetics and evolution of Hawaiian *Drosophila*. In M. Ashburner, H. L. Carson, and J. N. Thompson (eds), *The Genetics and Biology of Drosophila*, vol. 3b. Academic Press, New York, pp. 297-344.

CARSON, H. L., F. E. CLAYTON, and H. D. STALKER (1967). Karyotypic stability and speciation in Hawaiian *Drosophila*. *Proceedings of the National Academy of Sciences USA* **57**, 1280-5.

CARTMILL, M. (1981). Hypothesis testing and phylogenetic reconstruction. *Zeitschrift für Zoologische Systematik und Evolutionsforschung* **19**, 73-96.

CHARIG, A. J. (1981). Cladistics: a different point of view. *The Biologist* **28**, 19-20. (Reprinted 1982, in J. Maynard Smith (ed.), *Evolution Now*, Macmillan, London pp. 121-24.)

CHARIG, A. J. (1982). Systematics in biology: a fundamental comparison of some major schools of thought. In K. A. Joysey and A. E. Friday (eds), *Problems of Phylogenetic Reconstruction*. Academic Press, London, pp. 363-440.

CHARNOV, E. (1982). *The Theory of Sex-Allocation*. Princeton University Press, Princeton, New Jersey.

CLARK, R. B. (1964). *Dynamics in Metazoan Evolution*. Clarendon Press, Oxford.

COLEMAN, W. (1976). Morphology between type concept and descent theory. *Journal of the History of Medicine and Allied Sciences* **31**, 149-75.

COLLESS, D. H. (1966). A note on Wilson's consistency test for phylogenetic hypotheses. *Systematic Zoology* **15**, 358-9.

COLLESS, D. H. (1967). The phylogenetic fallacy. *Systematic Zoology* **16**, 288-95.

COOK, L. M., L. P. BROWER, and J. ALCOCK (1969). An attempt to verify mimetic advantage in a neotropical environment. *Evolution* **23**, 339-45.

COX, B. (1982). New branches for old. *Nature, London* **298**, 321.

CRACRAFT, J. (1974a). Phylogeny and evolution of the ratite birds. *Ibis* **116**, 494-521.

CRACRAFT, J. (1974b). Phylogenetic models and classification. *Systematic Zoology* **23**, 71-90.

CRACRAFT, J. (1979). Phylogenetic analysis, evolutionary models, and paleontology. In J. Cracraft and N. Eldredge (eds), *Phylogenetic Analysis and Paleontology*. Columbia University Press, New York, pp. 7-39.

CRACRAFT, J. (1981). The use of functional and adaptive criteria in phylogenetic systematics. *American Zoologist* **21**, 21-36.

CRACRAFT, J. (1983). Systematics, comparative biology, and the case against creationism. In L. R. Godfrey (ed.), *Scientists against Creationism*. Norton, New York, pp. 163-91.

CRICK, F. H. C. (1966). The genetic code - yesterday, today, and tomorrow. *Cold Spring Harbor Symposia on Quantitative Biology* **31**, 3-9.

CRICK, F. H. C. (1968). The origin of the genetic code. *Journal of Molecular Biology* **38**, 367-79.

CRISCI, J. V. and T. F. STUESSY (1980). Determining primitive character states for phylogenetic reconstruction. *Systematic Botany* **5**, 112-35.

CROMPTON, A. W. and F. A. JENKINS (1979). Origin of mammals. In J. A. Lillegraven, Z. Kielan-Jaworowska, and W. A. Clemens (eds), *Mesozoic Mammals*. University of California Press, Berkeley, pp. 59-73.

CROWSON, R. A. (1970). *Classification and Biology*. Heinemann, London.

CULLUM, J. and H. SAEDLER (1981). DNA rearrangements and evolution. In M. J. Carlile, J. F. Collins, and B. E. B. Moseley (eds), *Molecular and Cellular Aspects of Microbial Evolution*, 32nd Symposium of the Society for General Microbiology. Cambridge University Press, Cambridge, pp. 131-50.

CULVER, D. (1982). *Cave Life*. Harvard University Press, Cambridge, Massachusetts.

CUVIER, G. (1829). *Le Règne Animal*, nouvelle edition. Paris. (Translated 1834, as *The Animal Kingdom*, G. Henderson, London.)

DARWIN, C. (1859). *On the Origin of Species*. John Murray, London. (Reprinted 1969, J. W. Burrow (ed.), Penguin Books, Harmondsworth, which is the edition I cite in the text.)

DARWIN, F. ed. (1887). *The Life and Letters of Charles Darwin*, 3 vols. John Murray, London.

DAVY, J. (1981). Once upon a time ... *The Observer*, 16 August, pp. 19, 21.

DAWKINS, R. (1982). *The Extended Phenotype*. W. H. Freeman, Oxford.

DE BEER, G. R. (1931). *Embryology and Evolution*. Oxford University Press, Oxford.

DE BEER, G. R. (1940). Embryology and taxonomy. In J. Huxley (ed.), *The New Systematics*. Clarendon Press, Oxford, pp. 365-93.

DE BEER, G. R. (1956). The evolution of ratites. *Bulletin of the British Museum (Natural History)*, Zoological Series 4 57-76.

DE BEER, G. R. (1964). Ratites, phylogeny of the. In Sir A. Landsborough Thomson (ed.), *A New Dictionary of Birds*. Nelson, London, pp. 681-5.

DICKERSON, R. E. (1972). The structure and history of an ancient protein. *Scientific American* **226** (April), 58-72.

References

DOBZHANSKY, T. (1937). *Genetics and the Origin of Species.* Columbia University Press, New York.

DOBZHANSKY, T. (1970). *Genetics of the Evolutionary Process.* Columbia University Press, New York.

EDWARDS, A. W. F. (1983). Contribution to discussion after Felsenstein (1983b), q.v., pp. 262-3.

ELDREDGE, N. (1979). Cladism and common sense. In J. Cracraft and N. Eldredge (eds), *Phylogenetic Analysis and Paleontology.* Columbia University Press, New York, pp. 165-97.

ELDREDGE, N. and J. CRACRAFT. (1980). *Phylogenetic Patterns and the Evolutionary Process.* Columbia University Press, New York.

ELDREDGE, N. and I. TATTERSALL. (1977). Evolutionary models, phylogenetic reconstruction, and another look at hominid phylogeny. In J. Gray and A. J. Boucot (eds), *Historical Biogeography, Plate Tectonics and the Changing Environment.* Oregon State University Press, Corvallis, pp. 147-67.

ELTON, C. (1927). *Animal Ecology.* Sidgwick & Jackson, London.

EMDEN, F. van (1929). Über den Speciesbegriff vom Standpunkt der Larvensystematik. *3 Wanderversammlung Deutscher Entomologen in Geissen,* pp. 22-6, 47-56.

ENGELMANN, G. F. and E. O. WILEY (1977). The place of ancestor-descendant relationships in phylogeny reconstruction. *Systematic Zoology* **26**, 1-11.

FALKOW, S. (1975). *Infectious Multiple Disease Resistance.* Pion Press, London.

FARRIS, J. S. (1976). Phylogenetic classification of fossils with recent species. *Systematic Zoology* **25**, 271-82.

FARRIS, J. S. (1977). On the phenetic approach to vertebrate classification. In M. K. Hecht, P. C. Goody and B. M. Hecht (eds), *Major Patterns in Vertebrate Evolution.* Plenum Press, New York, pp. 823-50.

FARRIS, J. S. (1979). The information content of the phyletic system. *Systematic Zoology* **28**, 483-519.

FARRIS, J. S. (1983). The logical basis of phylogenetic analysis. In N. I. Platnick and V. A. Funk (eds), *Advances in Cladistics,* vol. 2. Columbia University Press, New York. (Reprinted 1984 in E. Sober (ed.), *Conceptual Issues in Evolutionary Biology,* MIT Press, Cambridge, Massachusetts.)

FELSENSTEIN, J. (1978). Cases in which parsimony or compatibility methods can be positively misleading. *Systematic Zoology* **27**, 401-10.

FELSENSTEIN, J. (1982). Numerical methods for inferring evolutionary trees. *Quarterly Review of Biology* **57**, 379-404.

FELSENSTEIN, J. (1983a). Parsimony in systematics: biological and statistical issues. *Annual Review of Ecology and Systematics* **14**, 313-33.

FELSENSTEIN, J. (1983b). Statistical inference of phylogenies (with discussion). *Journal of the Royal Statistical Society* A **146**, 246-72.

FINK, W. L. (1982). The conceptual relationship between ontogeny and phylogeny. *Paleobiology* **8**, 254-64.

FISHER, D. C. (1981). The role of functional analysis in phylogenetic inference: examples from the history of the Xiphosura. *American Zoologist* **21**, 47-62.

FOELIX, R. F. (1982). *Biology of Spiders*. Harvard University Press, Cambridge, Massachusetts.

FORD, E. B. (1941). Studies on the chemistry of pigments in the Lepidoptera, with special reference to their bearing on systematics. 1. The anthoxanthins. *Proceedings of the Royal Entomological Society of London* A **16**, 65-90.

FORD, E. B. (1942). Studies on the chemistry of pigments in the Lepidoptera, with special reference to their bearing on systematics. 2. Red pigments in the genus *Delias*. *Proceedings of the Royal Entomological Society of London* A **17**, 87-92.

FORD, E. B. (1944a). Studies on the chemistry of pigments in the Lepidoptera, with special reference to their bearing on systematics. 3. The red pigments of the Papilionidae. *Proceedings of the Royal Entomological Society of London* A **19**, 92-106.

FORD, E. B. (1944b). Studies on the chemistry of pigments in the Lepidoptera, with special reference to their bearing on systematics. 4. The classification of the Papilionidae. *Transactions of the Royal Entomological Society of London* **94**, 201-23.

FORD, E. B. (1947). Studies on the chemistry of pigments in the Lepidoptera, with special reference to their bearing on systematics. 5. *Pseudopontia paradoxa*. *Proceeding of the Royal Entomological Society of London* A **22**, 77-8.

FORD, E. B. (1975). *Ecological Genetics*, 4th edn. Chapman & Hall, London.

FOREY, P. L. (1982). Neonotological analysis versus palaeontological stories. In K. A. Joysey and A. E. Friday (eds). *Problems of Phylogenetic Reconstruction*. Academic Press, London, pp. 119-57.

FORTEY, R. A. and R. P. S. JEFFERIES (1982). Fossils and phylogeny - a compromise approach. In K. A. Joysey and A. E. Friday (eds), *Problems of Phylogenetic Reconstruction*. Academic Press, London, pp. 197-234.

FRIDAY, A. E. (1982). Parsimony, simplicity, and what actually happened. *Zoological Journal of the Linnean Society* **74**, 329-35.

FUNK, V. A. and D. R. BROOKS. (1981). National Science Foundation workshop on the theory and application of cladistic methodology. *Systematic Zoology* **30**, 491-8.

FÜRBRINGER, M. (1888). *Untersuchungen zur Morphologie und*

Systematik der Vogel, 2 vols. Verlag von T. J. van Holkema, Amsterdam.

GAFFNEY, E. S. (1979). An introduction to the logic of phylogeny reconstruction. In J. Cracraft and N. Eldredge (eds), *Phylogenetic Analysis and Paleontology.* Columbia University Press, New York, pp. 79-111.

GARDINER, B. G. (1982). Tetrapod classification. *Zoological Journal of the Linnean Society* **74,** 207-32.

GARDINER, B. G. (1983). Gnathostome vertebrae and the classification of the Amphibia. *Zoological Journal of the Linnean Society* **79,** 1-59.

GARDINER, B. G., P. JANVIER, C. PATTERSON, P. L. FOREY, P. H. GREENWOOD, R. S. MILES, and R. P. S. JEFFERIES (1979). The salmon, the lungfish, and the cow: a reply. *Nature, London* **277,** 175-6.

GARSTANG, W. (1922). The theory of recapitulation: a critical restatement of the biogenetic law. *Journal of the Linnean Society, Zoology* **35,** 81-101.

GARSTANG, W. (1928). The origin and evolution of larval forms. Presidential Address, Section D, *Reports of the British Association,* Glasgow, 23 pp.

GHISELIN, M. T. (1966). On psychologism in the logic of taxonomic controversies. *Systematic Zoology* **15,** 207-15.

GHISELIN, M. T. (1969). *The Triumph of the Darwinian Method.* University of California Press, Berkeley.

GHISELIN, M. T. (1976). Two Darwin's: history versus criticism. *Journal of the History of Biology* **9,** 121-32.

GHISELIN, M. T. (1984). 'Definition', 'character,' and other equivocal terms. *Systematic Zoology* **33,** 104-10.

GILLESPIE, N. C. (1979). *Charles Darwin and the Problem of Creation.* University of Chicago Press, Chicago.

GINGERICH, P. D. (1979). The stratophenetic approach to phylogeny reconstruction in vertebrate evolution. In J. Cracraft and N. Eldredge (eds), *Phylogenetic Analysis and Paleontology.* Columbia University Press, New York, pp. 41-77.

GOLDSCHMIDT, R. (1940). *The Material Basis of Evolution.* Yale University Press, New Haven, Connecticut.

GOODMAN, M. (1981). Decoding the pattern of protein evolution. *Progress in Biophysics and Molecular Biology* **38,** 105-64.

GOTO, H. E. (1982). *Animal Taxonomy.* Edward Arnold, London.

GOULD, S. J. (1977a). *Ontogeny and Phylogeny.* Harvard University Press, Cambridge, Massachusetts.

GOULD, S. J. (1977b). The telltale wishbone. *Natural History* **86** (November), 26-36. (Reprinted 1980, in S. J. Gould, *The Panda's Thumb,* Norton, New York, pp. 267-77.)

GOULD, S. J. (1980). Is a new and general theory of evolution emerging? *Paleobiology* **6,** 119-30.

Evolution and Classification

GRIFFITHS, G. D. C. (1972). *The Phylogenetic Classification of the Diptera Cyclorrhapha with Special Reference to the Structure of the Postabdomen.* Dr W. Junk, The Hague.

GROUCHY, J. de, C. TURLEAU, and C. FINAZ (1978). Chromosomal phylogeny of the primates. *Annual Review of Genetics* **12**, 289-328.

GÜNTHER, K. (1971). Abschliessende Zusammenfassung der Vorträge Diskossionen. In R. Siewing (ed.), *Methoden der Phylogenetik.* Erlanger Forschungen B, vol. 4, pp. 76-86.

GUTMANN, W. F. (1981). Relationships between invertebrate phyla based on functional-mechanical analysis of the hydrostatic skeleton. *American Zoologist* **21**, 63-81.

HALSTEAD, L. B. (1978). The cladistic revolution - can it make the grade? *Nature, London* **276**, 759-60.

HALSTEAD, L. B., E. I. WHITE, and G. T. MacINTYRE (1979). [Reply to Gardiner *et al.* (1979), q.v.] *Nature, London* **277**, 176.

HAMILTON, W. D. and M. ZUK (1982). Heritable true fitness and bright birds: a role for parasites. *Science, NY* **218**, 384-7.

HARDY, A. C. (1954). Escape from specialization. In J. S. Huxley, A. C. Hardy, and E. B. Ford (eds), *Evolution as a Process.* Allen & Unwin, London, pp. 122-42.

HARPER, C. W. (1976). Phylogenetic inference in paleontology. *Journal of Paleontology* **50**, 180-93.

HECHT, M. K. and J. L. EDWARDS (1977). The methodology of phylogenetic inference above the species level. In M. K. Hecht, P. C. Goody, and B. M. Hecht (eds), *Major Patterns of Vertebrate Evolution.* Plenum, New York, pp. 3-51.

HENDY, M. D. and D. PENNY (1984). Cladograms should be called trees. *Systematic Zoology* **33**, 245-7.

HENNIG, W. (1966). *Phylogenetic Systematics.* University of Illinois Press, Urbana.

HITCHING, F. (1982). *The Neck of the Giraffe.* Pan Books, London.

HOLMES, E. B. (1980). Reconsideration of some systematic concepts and terms. *Evolutionary Theory* **5**, 35-87.

HULL, D. L. (1965). The effect of essentialism on taxonomy. *British Journal for the Philosophy of Science* **15**, 314-26; **16**, 1-18.

HULL, D. L. (1967). Certainty and circularity in evolutionary taxonomy. *Evolution* **21**, 174-89.

HULL, D. L. (1968). The operational imperative: sense and nonsense in operationism. *Systematic Zoology* **17**, 438-57.

HULL, D. L. (1970). Contemporary systematic philosophies. *Annual Review of Ecology and Systematics* **1**, 19-54.

HULL, D. L. (1973). *Darwin and his Critics.* Harvard University Press, Cambridge, Massachusetts.

HULL, D. L. (1979). The limits of cladism. *Systematic Zoology* **28**, 416-40.

References

HULL, D. L. (1980). [Review of J. Cracraft and N. Eldredge (eds), *Phylogenetic Analysis and Paleontology.*] *Paleobiology* **6**, 131-6.

HULL, D. L. (1983). Karl Popper and Plato's metaphor. In N. Platnick and V. Funk (eds), *Advances in Cladistics*, vol. 2. Columbia University Press, New York, pp. 177-89.

HUMPHRIES, C. J. (1983). [Review of K. A. Joysey and A. E. Friday (eds), *Problems of Phylogenetic Reconstruction.*] *Systematic Zoology* **32**, 302-10.

HUXLEY, J. S. ed. (1940). *The New Systematics*. Clarendon Press, Oxford.

HUXLEY, J. S. (1942). *Evolution: The Modern Synthesis*. Allen & Unwin, London.

HUXLEY, J. S. (1958). Evolutionary processes and taxonomy with special reference to grades. *Uppsala Universitets Arsskrift* **1958**, pp. 21-39.

HYMAN, L. H. (1951). *The Invertebrates: Acanthocephala, Aschelminthes, and Entoprocta*. McGraw-Hill, New York.

JANVIER, P. (1984). Cladistics: theory, purpose, and evolutionary implications. In J. W. Pollard (ed.), *Evolutionary Theory: Paths into the Future*. Wiley, Chichester, pp. 39-75.

JARVIK, E. (1981). [Review of Rosen *et al.* (1981), q.v.] *Systematic Zoology* **30**, 378-84.

JOHNSON, L. A. S. (1970). Rainbow's end: the quest for an optimal taxonomy. *Systematic Zoology* **19**, 203-39.

KEMP, T. S. (1982a). *Mammal-like Reptiles and the Origin of the Mammals*. Academic Press, London.

KEMP, T. S. (1982b). The reptiles that became mammals. *New Scientist* **93**, 581-4. (Reprinted 1982, in J. J. Cherfas (ed.), *Darwin Up to Date*, New Scientist, London, pp. 31-4.)

KEMP, T. S. (1983). The relationships of mammals. *Zoological Journal of the Linnean Society* **77**, 353-84.

KIMURA, M. (1968). Evolutionary rate at the molecular level. *Nature, London* **217**, 624-6.

KIMURA, M. (1983). *The Neutral Theory of Molecular Evolution*. Cambridge University Press, Cambridge.

KOTTLER, M. J. (1978). Charles Darwin's biological species concept and the theory of geographic speciation: the transmutation notebooks. *Annals of Science* **35**, 275-97.

LACK, D. (1966). *Population Studies of Birds*. Clarendon Press, Oxford.

LANDE, R. (1978). Evolutionary mechanisms of limb loss in tetrapods. *Evolution* **32**, 73-92.

LANKESTER, E. R. (1873). On the primitive cell-layers of the embryo as the basis of genealogical classification of animals, and on the origin of vascular and lymph systems. *Annals and Magazine of Natural History*, ser. 4 **11**, 321-38.

Evolution and Classification

LE GROS CLARK, W. E. (1959). *The Antecedents of Man*. Edinburgh University Press, Edinburgh.

LEITH, B. (1981). Are the reports of Darwin's death exaggerated? *The Listener* **106** (8 October), 390-2.

LEITH, B. (1982). *The Descent of Darwin*. Collins, London.

LE QUESNE, W. J. (1969). A method of selection of characters in numerical taxonomy. *Systematic Zoology* **18**, 201-5.

LE QUESNE, W. J. (1982). Compatibility analysis and its application. *Zoological Journal of the Linnean Society* **74**, 267-75.

LEWIN, R. (1982a). Where is the science in creation science? *Science, NY* **215**, 142-6.

LEWIN, R. (1982b). Judge's ruling hits hard at creationism. *Science, NY* **215**, 381-4.

LEWIN, R. (1982c). Can genes jump between eukaryotic species? *Science, NY* **217**, 42-3.

LORENZ, K. (1941). Vergleichende Bewegungsstudien an Anatiden. *Journal für Ornithologie* **79**, 194-294. (Translated by R. Martin in K. Lorenz, 1971, *Studies in Animal and Human Behaviour*, vol. 2, Methuen, London, pp. 14-114.

LYELL, C. (1830-3). *The Principles of Geology*, 3 vols. John Murray, London.

McGOWAN, C. (1984). Evolutionary relationships of ratites and carinates: evidence from ontogeny of the tarsus. *Nature, London* **307**, 733-5.

MacLEAY, W. S. (1825). Remarks on the identity of certain laws which have been lately observed to regulate the natural distribution of insects and fungi. *Transactions of the Linnean Society of London* **14**, 46-68.

McNEILL, J. (1982). Phylogenetic reconstruction and phenetic taxonomy. *Zoological Journal of the Linnean Society* **74**, 337-44.

McNEILL, J. (1983). The future of numerical methods in plant systematics: a personal prospect. In J. Felsenstein (ed.), *Numerical Taxonomy*. Springer-Verlag, Berlin, pp. 47-52.

MADDISON, W. P., M. J. DONOGHUE and D. R. MADDISON (1984). Outgroup analysis and parsimony. *Systematic Zoology* **33**, 83-103.

MANTON, S. (1977). *The Arthropoda*. Clarendon Press, Oxford.

MARTIN, R. (1981). Phylogenetic reconstruction versus classification: the case for clear demarcation. *The Biologist* **28**, 127-32.

MAYNARD SMITH, J. (1978). Optimization theory in evolution. *Annual Review of Ecology and Systematics* **9**, 31-56.

MAYNARD SMITH, J. (1982). *Evolution and the Theory of Games*. Cambridge University Press, Cambridge.

MAYR, E. (1942). *Systematics and the Origin of Species*. Columbia University Press, New York.

MAYR, E. (1963). *Animal Species and Evolution*. Harvard University Press, Cambridge, Massachusetts.

References

MAYR, E. (1969). *Principles of Systematic Zoology*. McGraw-Hill, New York.

MAYR, E. (1974). Cladistic analysis or cladistic classification? *Zeitschrift für Zoologische Systematik und Evolutionsforschung* 12, 94-128. (Reprinted in Mayr (1976), q.v., pp. 433-76.)

MAYR, E. (1976). *Evolution and the Diversity of Life*. Harvard University Press, Cambridge, Massachusetts.

MAYR, E. (1980). G. G. Simpson. In E. Mayr and W. B. Provine (eds), *The Evolutionary Synthesis*. Harvard University Press, Cambridge, Massachusetts, pp. 452-63.

MAYR, E. (1981). Biological classification: toward a synthesis of opposing methodologies. *Science, NY* 214, 510-16.

MAYR, E., E. G. LINSLEY, and R. L. USINGER (1953). *Methods and Principles of Systematic Zoology*. McGraw-Hill, New York.

MEACHAM, C. A. (1984). The role of hypothesized direction of characters in the estimation of evolutionary history. *Taxon* 33, 26-38.

MICKEVICH, M. F. (1978). Taxonomic congruence. *Systematic Zoology* 27, 143-58.

MORTON, E. S. (1977). On the occurrence and significance of motivation-structural rules in some bird and mammal sounds. *American Naturalist* 111, 855-69.

NELSON, G. (1972). Comments on Hennig's 'Phylogenetic Systematics' and its influence on ichthyology. *Systematic Zoology* 21, 364-74.

NELSON, G. (1973). 'Monophyly again?' - a reply to P. D. Ashlock. *Systematic Zoology* 22, 310-12.

NELSON, G. (1978). Ontogeny, phylogeny, paleontology, and the biogenetic law. *Systematic Zoology* 27, 324-45.

NELSON, G. (1979). Cladistic analysis and synthesis: principles and definitions, with a historical note on Adanson's *Familles des Plantes* (1763-1764). *Systematic Zoology* 28, 1-21.

NELSON, G. and N. I. PLATNICK (1980). Multiple branches in cladograms: two interpretations. *Systematic Zoology* 29, 86-91.

NELSON, G. and N. I. PLATNICK (1981). *Systematics and Biogeography*. Columbia University Press, New York.

NELSON, G. and N. I. PLATNICK (1984). Systematics and evolution. In M.-W. Ho and P. T. Saunders (eds), *Beyond Neo-Darwinism*. Academic Press, London, pp. 143-58.

NIELSEN, C. (1971). Entoproct life-cycles and the entoproct/ectoproct relationship. *Ophelia* 9, 209-341.

NIELSEN, C. (1977). Phylogenetic considerations: the protostomian relations. In R. M. Woollacott and R. L. Zimmer (eds), *Biology of Bryozoans*. Academic Press, New York, pp. 519-34.

OSPOVAT, D. (1981). *The Development of Darwin's Theory*. Cambridge University Press, Cambridge.

OWEN, R. (1848). *On the Archetype and Homologies of the Vertebrate Skeleton*. For the Author, London.

PANCHEN, A. L. (1982). The use of parsimony in testing phylogenetic hypotheses. *Zoological Journal of the Linnean Society* 74, 305-28.
PARSONS, T. S. and E. E. WILLIAMS (1963). The relationships of the modern Amphibia: a re-examination. *Quarterly Review of Biology* 38, 26-53.
PASSMORE, J. (1970). *The Perfectibility of Man*. Duckworth, London.
PATTERSON, C. (1978). Verifiability in systematics. *Systematic Zoology* 27, 218-22.
PATTERSON, C. (1980). Cladistics. *Biologist* 27, 234-40. (Reprinted 1982, in J. Maynard Smith (ed.), *Evolution Now*, Macmillan, London, pp. 110-20. Page references in the text are to this reprint.)
PATTERSON, C. (1981). Significance of fossils in determining evolutionary relationships. *Annual Review of Ecology and Systematics* 12, 195-223.
PATTERSON, C. (1982a). Cladistics and classification. *New Scientist* 94, 303-6. (Reprinted 1982, in J. J. Cherfas (ed.), *Darwin Up to Date*, New Scientist, London, pp. 35-8. Page references in the text are to this reprint.)
PATTERSON, C. (1982b). Morphological characters and homology. In K. A. Joysey and A. E. Friday (eds), *Problems of Phylogenetic Reconstruction*. Academic Press, London, pp. 21-74.
PATTERSON, C. (1983). How does phylogeny differ from ontogeny? In B. C. Goodwin, N. Holder, and C. C. Wylie (eds), *Development and Evolution*, 6th Symposium of the British Society for Developmental Biology. Cambridge University Press, Cambridge, pp. 1-31.
PAUL, C. R. C. (1982). The adequacy of the fossil record. In K. A. Joysey and A. E. Friday (eds), *Problems of Phylogenetic Reconstruction*. Academic Press, London, pp. 75-117.
PELLEGRIN, P. (1982). *La classification des animaux chez Aristote*. Les Belles Lettres, Paris.
PENNY, D., L. R. FOULDS, and M. D. HENDY (1982). Testing the theory of evolution by comparing phylogenetic trees constructed from five different protein sequences. *Nature, London* 297, 197-200.
PLATNICK, N. I. (1977). Cladograms, phylogenetic trees, and hypothesis testing. *Systematic Zoology* 26, 438-42.
PLATNICK, N. I. (1979). Philosophy and the transformation of cladistics. *Systematic Zoology* 28, 537-46.
PLATNICK, N. I. (1982). Defining characters and evolutionary groups. *Systematic Zoology* 31, 282-4.
POPPER, K. (1980). [Letter to the editor.] *New Scientist* 87, 611.
PROTHERO, D. R. and D. B. LAZERUS (1980). Planktonic microfossils and the recognition of ancestors. *Systematic Zoology* 29, 119-29.
REHBOCK, P. F. (1983). *The Philosophical Naturalists*. University of Wisconsin Press, Madison, Wisconsin.

References

RIDLEY, M. (1982). Coadaptation and the inadequacy of natural selection. *British Journal for the History of Science* **15**, 45-68.

RIDLEY, M. (1983). *The Explanation of Organic Diversity.* Clarendon Press, Oxford.

RIDLEY, M. (1986). Embryology and classical zoology in Great Britain. In T. Horder (ed.), *A Critical History of Embryology.* Cambridge University Press, Cambridge.

RIEPPEL, O. (1980). Homology, a deductive concept? *Zeitschrift für Zoologische Systematik und Evolutionsforschung* **18**, 315-19.

RIEPPEL, O. (1984). Atomism, transformism, and the fossil record. *Zoological Journal of the Linnean Society* **82**, 17-32.

ROHLF, F. J. (1974). Methods of comparing classifications. *Annual Review of Ecology and Systematics* **5**, 101-13.

ROHLF, F. J. and R. R. SOKAL (1981). Comparing numerical taxonomic studies. *Systematic Zoology* **27**, 143-58.

ROSEN, D. E. (1982). Do current theories of evolution satisfy the basic requirements of explanation? *Systematic Zoology* **31**, 76-85.

ROSEN, D. E., P. L. FOREY, B. G. GARDINER, and C. PATTERSON (1981). Lungfishes, tetrapods, paleontology, and plesiomorphy. *Bulletin of the American Museum of Natural History* **167**, 159-276.

RUSSELL, E. S. (1916). *Form and Function.* John Murray, London.

SCHAEFFER, B., M. K. HECHT, and N. ELDREDGE (1972). Phylogeny and paleontology. *Evolutionary Biology* **6**, 31-46.

SCHINDEWOLF, O. (1968). Homologie und Taxonomie. *Acta Biotheoretica* **18**, 235-83.

SCHMIDT, R. S. (1960). Predator behaviour and the perfection of incipient mimetic resemblances. *Behaviour* **16**, 149-58.

SEDGWICK, A. (1894). On the law of development commonly known as Von Baer's Law; and on the significance of ancestral rudiments in embryonic development. *Quarterly Review of Microscopical Science* **36**, 35-52.

SEDGWICK, A. (1909). The influence of Darwin on the study of animal embryology. In A. Seward (ed.), *Darwin and Modern Science.* Cambridge University Press, Cambridge, pp. 171-84.

SEVERTZOFF, A. N. (1929). Directions of evolution. *Acta Zoologica* **10**, 59-141.

SILBERGLIED, R. E., A. AIELLO, and D. M. WINDSOR (1980). Disruptive coloration in butterflies: lack of support in *Anartia fatima*. *Science, NY* **209**, 617-19.

SIMPSON, G. G. (1953). *The Major Features of Evolution.* Columbia University Press, New York.

SIMPSON, G. G. (1959a). Mesozoic mammals and the polyphyletic origin of mammals. *Evolution* **13**, 405-14.

SIMPSON, G. G. (1959b). The nature and origin of supraspecific taxa. *Cold Spring Harbor Symposia in Quantitative Biology* **24**, 255-71.

SIMPSON, G. G. (1961). *Principles of Animal Taxonomy*. Columbia University Press, New York.

SIMPSON, G. G. (1975). Recent advances in the methods of phylogenetic inference. In W. P. Lucket and F. S. Szalay (eds), *Phylogeny of the Primates: a Multidisciplinary Approach*. Plenum Press, New York, pp. 3-19.

SIMPSON, G. G. (1978). *Concession to the Improbable*. Yale University Press, New Haven, Connecticut.

SIMPSON, G. G. (1980). *Splendid Isolation*. Yale University Press, New Haven, Connecticut.

SKINNER, Q. (1965). History and ideology in the English revolution. *Historical Journal* 8, 151-78.

SLOAN, P. R. (1972). John Locke, John Ray, and the problem of the natural system. *Journal for the History of Biology* 5, 1-53.

SNEATH, P. H. A. (1982). [Review of Nelson and Platnick (1981), q.v.] *Systematic Zoology* 31, 208-17.

SNEATH, P. H. A. and R. R. SOKAL (1973). *Numerical Taxonomy*. W. H. Freeman, San Francisco.

SOKAL, R. R. (1973). [Contribution to discussion.] *Systematic Zoology* 22, 399-400.

SOKAL, R. R. (1975). Mayr on cladism - and his critics. *Systematic Zoology* 24, 257-62.

SOKAL, R. R. and P. H. A. SNEATH (1963). *The Principles of Numerical Taxonomy*. W. H. Freeman, San Francisco.

SPRAGUE, T. A. (1940). Taxonomic botany, with special reference to the angiosperms. In J. S. Huxley, ed. (1940), q.v., pp. 435-54.

STACKERBRANDT, E. and C. R. WOESE (1981). The evolution of prokaryotes. In H. J. Carlile, J. F. Collins, and B. E. B. Moseley (eds), *Molecular and Cellular Aspects of Microbial Evolution*, 32nd Symposium of the Society for General Microbiology. Cambridge University Press, Cambridge, pp. 1-31.

STALKER, H. D. (1972). Intergroup phylogenies in *Drosophila* as determined by comparisons of salivary banding patterns. *Genetics* 70, 457-74.

STEVENS, P. F. (1980). Evolutionary polarity of character states. *Annual Review of Ecology and Systematics* 11, 333-58.

STEVENS, P. F. (1983). Report of third annual Willi Hennig Society Meeting. *Systematic Zoology* 32, 285-91.

SUTHERLAND, L. D. and G. E. PARKER (1981). Evolution? Prominent scientist reconsiders. *Impact,* no. 108, 4 pp.

SWAINSON, W. [1835]. *A Treatise on the Geography and Classification of Animals*. Longman, Brown, Green, &c., London.

SZALAY, F. S. (1981). Ancestors, descendants, sister groups, and testing of phylogenetic hypotheses. *Systematic Zoology* 26, 12-18.

SZALAY, F. S. (1981). Functional analysis and the practice of the phylogenetic method as reflected by some mammalian studies. *American Zoologist* 21, 37-45.

References

TEFFORDS, M. R., J. G. STERNBURG, and G. P. WALDBAUER (1979). Batesian mimicry: field demonstration of the survival value of pipevine swallowtail and monarch color patterns. *Evolution* **33**, 275-86.

THOMPSON, J. V. (1830). On the cirripedes or barnacles; demonstrating their deceptive character; the extraordinary metamorphoses they undergo, and the class of animals to which they indisputably belong. *Zoological Researches,* memoir 4, King and Riding, Cork.

THOMPSON, J. V. (1835). Discovery of the metamorphosis in the second type of cirripedes, viz., the Lepades, completing the natural history of these singular animals, and confirming their affinity with the Crustacea. *Philosophical Transactions of the Royal Society of London,* pp. 335-9.

THOMSON, K. S. (1982). The meanings of evolution. *American Scientist* **70**, 529-31.

TREVOR-ROPER, H. R. (1966). George Buchanan and the Ancient Scottish Constitution. *English Historical Review,* Supplement no. 3, 53 pp.

VAN VALEN, L. (1978). Why not to be a cladist. *Evolutionary Theory* **3**, 285-99.

VON EHRENSTEIN, G. and F. LIPMANN (1962). Experiments on hemoglobin biosynthesis. *Proceedings of the National Academy of Sciences USA* **47**, 941-50.

WARBURTON, F. E. (1967). The purposes of classification. *Systematic Zoology* **16**, 241-5.

WASSERMAN, M. (1954). Cytological studies of the *repleta* group. *University of Texas Publications* no. 5422, 130-52.

WASSERMAN, M. (1963). Cytology and phylogeny of Drosophila. *American Naturalist* **97**, 333-62.

WEINSTEIN, I. B. (1963). Comparative studies on the genetic code. *Cold Spring Harbor Symposia on Quantitative Biology,* **28**, 579-80.

WEINSTEIN, I. B. and A. N. SCHECHTER (1962). Polyuridylic acid stimulation of phenylalanine incorporation in animal cell extracts. *Proceedings of the National Academy of Sciences USA* **48**, 1686-91.

WEISMANN, A. (1882). *Studies in the Theory of Descent,* translated by R. Meldola. Sampson Low, London.

WILEY, E. O. (1978). The evolutionary species concept reconsidered. *Systematic Zoology* **27**, 17-26.

WILEY, E. O. (1979). An annotated Linnean hierarchy. with comments on natural taxa and competing systems. *Systematic Zoology* **28**, 308-37.

WILEY, E. O. (1981). *Phylogenetics.* Wiley, New York.

WILLIAMSON, M. (1981). *Island Populations.* Oxford University Press, Oxford.

WILSON, E. O. (1965). A consistency test for phylogenies based on contemporaneous species. *Systematic Zoology* **14**, 214-20.

WILSON, E. O. (1967). The validity of the 'consistency test' for phylogenetic hypotheses. *Systematic Zoology* **16**, 104.

YEO, R. (1979). William Whewell, natural theology, and the philosophy of science in mid nineteenth century Britain. *Annals of Science* **36**, 493-516.

YOUNG, D. A. (1983). Botanical cladistics: nothing new? *Taxon* **32**, 275-8.

ZANGERL, R. (1948). The methods of comparative anatomy and its contribution to the study of evolution. *Evolution* **2**, 351-74.

Glossary

The glossary concentrates on two kinds of word: those that are used in ordinary language, but have peculiar meanings in taxonomy; and ambiguous technical terms. I have taken the opportunity to explain which meanings I have chosen to follow. I have tried to explain, rather than simply define, the terms, as the glossary is intended mainly for readers who may not already be familiar with the language of philosophical classification. Italics indicate cross-references.

accident See *substance*.

adaptation A property of an organism well designed to carry out some function in the life of the organism. Eyes, for instance, are adaptations for seeing; optical analysis of the lens, the shape of the eye, and the photoreceptive properties of the retinal cells, all reveal that the eye is designed to produce visual images. It requires study to discover what any particular character of an organism is an adaptation for, and the discovery is easier in some cases than in others.

analogy A similarity, of form or structure, between two species not shared with their nearest common ancestor. The analogous structure has evolved independently in the two species; it is due to convergent evolution. In pre-Darwinian biology, analogous structures were those that superficially looked similar but on deeper study were found to differ. The wings of birds and bats are analogous organs: in the modern evolutionary sense because the common ancestor of birds and bats lacked wings; in the pre-Darwinian sense because, although they look like each other, their supporting skeletons differ. *Homology* is the opposite of analogy.

ancestral character state The evolutionarily earlier state of a character, relative to its derived state. As a character evolves through several states, each state is ancestral relative to the later states, and derived relative to the earlier states. Ancestral character states have to be distinguished from derived ones in *cladism*, which uses for this

Evolution and Classification

purpose such techniques as *outgroup comparision* and the *embryological criterion*. Taxonomists of the cladistic school often call ancestral character states 'plesiomorphies' and derived ones 'apomorphies'. Primitive is another synonym of ancestral, and 'advanced' of derived. 'Ancestral character state' has elsewhere also been used to mean the state of a character in an ancestral (or *stem*) species, but this is not the meaning used in this book (p. 54).

Aristotelian classification See *essentialist classification* and *substance and accident*.

artificial classification The term (like its opposite, *natural classification*) has many meanings; in this book I have picked a *phenetic* meaning. A classificatory group will be defined by certain characters, called 'defining' characters; in an artificial classification, the members of a group resemble one another in their defining characters (as they must, by definition) but not in their non-defining characters. With respect to the characters not used in the classification, the members of a group are uncorrelated.

cladism Classification by means of shared *derived* characters. Shared derived characters indicate *monophyletic groups*, and cladism is therefore purely *phylogenetic classification*. It classifies species into higher groups only by the recency of their common ancestry, not by their phenetic similarity. The same procedures, however, can be applied without making any theoretical assumptions (such as that the techniques indicate phylogenetic groups), which is the practice of *transformed cladism*. In this book, when the term 'cladism' is not accompanied by an epithet, I am talking about the evolutionary version of cladism, which I often call Hennigian cladism (after its inventor, Hennig). The groups of a cladistic classification are sometimes called clades (to contrast with *grades*), and the classification itself a *cladogram*.

cladogram A cladistic classification. This is the only meaning that I use in this book, and is the normal meaning in the modern literature; but the term has sometimes been used as a synonym of phylogenetic tree.

cluster statistic A statistical procedure for arranging units (such as organism, or species) into groups ('clusters') according to measurements made in those units. It is a *numerical* method of classification. There are a variety of kinds of cluster statistic, which differ according to the exact method by which they form groups.

coelom A body cavity found in many (but not all) kinds of multicellular animals; it contains a fluid – the coelomic fluid – and is lined by a cellular epithelium called the peritoneum. Of classificatory importance because it was believed to indicate a major phylogenetic

Glossary

group of animals; the coeloms of different groups were widely thought to be *homologous*, but this is now doubted.

congruence Non-contradictory taxonomic distribution of different characters. More exactly, two characters are said to be congruent if they fall into the same classificatory hierarchy. Two characters that have the same taxonomic distribution are trivially congruent; but characters are also called congruent if they have different distributions but fit the same classification. Thus, if, among five species, one character was found in species 3, 4, and 5 and another in species 3 and 4, they would be congruent because the group of species 3 and 4 may be a subgroup within the group of species 3, 4, and 5. But if one character was found in species 3, 4, and 5 and another in 1 and 5 they would be non-congruent, because the group of species 1 and 5 cannot be a subgroup of species 3, 4, and 5. I generally use the words 'agreement' and 'disagreement' (or 'contradiction') for congruence and non-congruence. Different characters may be non-congruent because they are convergent (see *analogy*) or because they are *ancestral* (p. 55); correctly identified derived characters should be perfectly congruent. Practically identified derived characters are often non-congruent, which makes for a problem in *cladism,* of which characters to use. The principle of *parsimony* provides one solution.

convergence See *analogy.*

derived character state The evolutionarily later state of a character, relative to its *ancestral* state. As a character evolves through several states, each state is derived relative to the state that preceded it (and derived relative to the one that follows it). In *cladism,* derived character states are sometimes called 'apomorphies' and a derived character state that is shared between more than one species a 'synapomorphy'. 'Advanced' is another synonym of derived. Shared derived characters are used to define the groups of a cladistic classification.

dichotomous classification Classification in which each larger group is divided into two smaller groups. Whether classification should be dichotomous and, if so, why are questions discussed in Chapters 6 and 10.

embryological criterion Embryological technique to distinguish *analogies* from *homologies* and *ancestral* from *derived* character states. As a technique of recognizing homologies it has a trivial and an interesting form. Trivially, a character cannot be homologous in two species unless its whole embryological development as well as its end form are ancestrally similar. The interesting form of the criterion, which is quite possibly wrong, asserts that character similarities between individuals of different species in early developmental states are more likely to be homologous than are similarities among adults.

Evolution and Classification

If correct, when characters from different stages of development are non-*congruent*, the developmentally earlier ones would be more reliable. There is also an embryological criterion by which to distinguish ancestral from derived character states (p. 66): according to this embryological criterion, the successive developmental stages of a character are also its successively more derived states: characters develop as well as evolve from more ancestral to more derived states. This last-mentioned criterion relies on the truth of *von Baer's law*, in an evolutionary interpretation.

essentialist classification Classification in which the groups are defined by particular characters, called 'essential' characters or 'essences'. Evolutionary classifications are not essentialist, because the groups are defined by the order of *phylogenetic* branching: characters do indicate the order of branching, and contingently may define phylogenetic groups; but they do not determine the order of branching. If a character that happens to be distributed in a monophyletic group undergoes an evolutionary change it will not alter the pattern of branching, it will merely cease to define a phylogenetic group. (See also *substance and accident*.)

eukaryotes See *prokaryotes*.

evolutionary taxonomy In terms of the three main kinds of groups of species – *monophyletic, paraphyletic,* and *polyphyletic groups* – evolutionary taxonomy aims to classify species into monophyletic or paraphyletic, not polyphyletic, groups. It is historically the most important post-Darwinian school of classification, and one of the three main schools discussed in this book (Chapter 2).

functional criterion Technique to distinguish *homologies* from *analogies,* and *ancestral* from *derived* character states, by means of the theory of natural selection. A character in two species cannot be homologous if the intermediate stages of that character between two species would have been non-functional – in which case they would have been prevented by natural selection. Ultimately, of course, all characters must have been connected by functional intermediate stages, which is why I wrote 'intermediate stages of that character' in the last sentence. The functional criterion requires that the intermediate stages be functional, and recognizably be forms of the same character: if the intermediate stages would have been a different character, the end points are analogies. If the theory of natural selection suggests that one direction of evolution between two character states is more likely than the other, the state which can more easily be evolved from it is more likely to be ancestral, and the other state derived.

genetic code Genes instruct the development of bodies by means of a code, written in triplets of molecules called bases. Each triplet

Glossary

encodes a particular amino acid; and the proteins that build bodies owe their properites to the sequence of amino acids of which they are made. The genetic code is universal, but arbitrary, which suggests a unique origin of life.

grade Group of species defined by a shared level of *adaptation*.

Hennigian classification See *cladism*.

holophyletic Group of species made up of all the descendant species from a common ancestral (*stem*) species. In this book, a dispensable (and dispensed-with) synonym of *monophyletic*.

homology A similarity, of form or structure, between two species shared from their common ancestor. The pentadactyl limb of humans and chimps, for example, was almost certainly present in their common ancestor, and is therefore homologous in the two species. Some *cladists* (p. 30) have suggested a more restricted use, according to which homology would mean a shared *derived character*; but I have not followed their suggestion in this book. Homology here implies shared *ancestral* characters as well as shared derived ones; this is the normal post-Darwinian meaning. Before Darwin, homology meant 'the same' character in different species: the pentadactyl limb would have been called homologous because of its formal identity in different species, not because it was shared from a common ancestor. *Analogy* is the opposite term.

idealism Philosophical doctrine that nature has an 'ideal' plan. In classification it means that the species of a classificatory group all have the same plan. It is a pre-evolutionary doctrine, and has more than one version (p. 106).

inversion, chromosomal A chromosomal variant in which a sequences of genes has been turned round. If we symbolize genes by letters of the alphabet, the chromosomes... ABCDEFGHI... and ...ABGFEDCHI... contain inversions of each other.

Linnaean hierarchy Traditional method of classifying species into a hierarchy of increasingly inclusive groups. The main categoric levels of the hierarchy, in ascending order, are: species, genus, family, order, class, phylum, kingdom. Other categories may be added, at any level, to accommodate the forms of particular groups. Species are customarily referred to by a Linnaean binomial, made up of its generic and specific name, as: *Homo sapiens* (genus *Homo*, species *sapiens*).

minimum evolution tree Reconstructed evolutionary *tree* (or *phylogeny*) consistent with the smallest number of evolutionary changes in the observed character states. Evolutionary trees are reconstructed using observations on the distribution of character states among species; the minimum evolution tree minimizes the number of changes required to have taken place in the observed characters. Its reconstruction relies on the principle of *parsimony*.

molecular clock The theory that molecular evolution proceeds at a fairly constant rate through time. The amount of molecular difference between two species is therefore proportional to the time since they shared a common ancestor.

monophyletic group In this book, a monophyletic group is one containing all the known descendants of an ancestral (*stem*) species. It has also been used in a broader sense, to mean a group of species that share a common ancestor that would be classified as a member of the group; the broader sense includes all monophyletic groups in the narrower sense, as well as groups containing only those descendants of a common ancestor that resemble the ancestor. Reptilia, for instance, are monophyletic in the broad, but not the narrow, sense: birds and mammals are descended from the common ancestor of the Reptilia, but are not included in the group Reptilia. In this book, Reptilia are called a *paraphyletic group,* to distinguish them from 'true' monophyletic groups. Other meanings of monophyly also exist (p. 30).

morphology The study of the forms of organisms; also used to refer to the form itself of an organism.

natural classification Classificatory groups are defined by certain characters, called 'defining' characters; in a natural group, the members of the group resemble one another for non-defining characters as well as for the defining character. This is not the only meaning of what is perhaps the most variously used term in taxonomy; but it is the meaning consistently used in this book, with rare and obvious exceptions. *Artificial classification* is the opposite of natural classification.

numerical classification In general, classification using quantitatively measured characters. In practice the term has become identified with a particular school of classification, the *phenetics* of Sokal and Sneath (1963). I therefore often use it as a synonym of phenetic classification. Of course, numerical techniques are used in *phylogenetic classification* too; but for historical reasons (see quotation on p. 36), the term numerical classification is more often used in phenetic than in phylogenetic classification.

objective character choice, objective classification Choice of characters, and therefore classification, according to an unambiguous and realistic theoretical principle. *Subjective character choice* is the alternative.

operationism Positivist philosophical doctrine that scientific terms mean no more than the operations carried out to measure them. If temperature were measured by a thermometer, for example, the operational meaning of temperature would be the height of a coloured liquid in a tube. Many scientists, however, implicitly distinguish theoretical terms from the evidence used to indicate them; *phylogenetic* taxonomists, for example, infer phylogenetic relations by means of similarities in certain kinds of characters (*homologies* and shared *derived* characters). Operationists deny the distinction of theory and evidence. The philosophy tends to be popular among critics of phylogenetic classification. (See especially Hull 1968.)

outgroup comparison Technique to distinguish *ancestral* from *derived* character states. If two states of a character are found in a group of species, the state possessed by a closely related species (called the outgroup) is inferred, by this technique, to be ancestral.

parsimony Principle that evolution proceeded by a smaller, rather than a larger, number of events. Humans and chimps, for example, share (among many other things) a characteristic structure called the eye. The inference that the common ancestor of the two species likewise possessed an eye, rather than that the organ evolved independently in the two species, relies on the principle of parsimony. Phylogenetic reconstruction relies on the principle, often in the extreme form of 'maximum parsimony', according to which evolution took the fewest possible steps given the distribution of character states among species. The *minimum evolution tree* is one application of the principle of maximum parsimony.

paraphyletic group A group made up of only those descendants of a common ancestor that resemble it phenetically. The group is so defined that if the common ancestor of it were discovered, it would be included in the group. Contrast *monophyletic* and *polyphyletic groups*.

pattern cladism See *transformed cladism*.

pattern–process distinction Pattern here means the distribution of taxonomic characters among species in nature, and the process is the natural mechanism which caused the characters to be distributed in the pattern they are. The 'pattern' is really a classification and the 'process' evolution by natural selection. Many *cladists,* particularly *transformed cladists,* maintain that the recognition of the 'pattern' of nature (that is, classification) should be kept separate from, and carried out before, any attempt to explain it by a theory of 'process' (that is, evolution). The theory of evolution, in other words, should be kept out of classification. When I talk about the pattern–process distinction in this book, I generally have the transformed cladistic application of it in mind.

phenetic classification The phenotype of an organism is simply its observed set of characteristics, and phenetic classification is classifica-

Evolution and Classification

tion of species solely according to the similarity of their phenotypes. One of the two main principles of classification, together with *phylogenetic classification*.

phylogenetic classification Classification solely by the *phylogeny* of species; species are grouped with those other species that they share their most recent common ancestor with. One of the two main principles of classification, together with *phenetic classification*. It is the principle applied by the school of *cladism*.

phylogeny The phylogeny (or phylogenetic *tree*) of a set of species shows their ancestral relations: a species is connected most closely to the other species it shares its most recent common ancestor with. Phylogenies are often drawn in two dimensions, with the species across the top and lines connecting them down the page; the vertical axis may then represent time, with the age of the common ancestor of two species being the vertical distance to the point where the two species are joined.

polyphyletic group A group of species defined in such a way that its most recent common ancestor would not be included in the group; like a *paraphyletic group,* it contains only some of the descendants of a *stem species,* but the stem species is excluded from polyphyletic groups, whereas it is included in paraphyletic ones. The defining characters of a polyphyletic group must have evolved independently in the member species; they are *analogies*. Polyphyletic groups are admitted in *phenetic classification,* but not in *cladism* or *evolutionary classification*.

process See *pattern-process distinction.*

prokaryotes A fundamental division of living things. All species can be classified either as eukaryotes (such as higher plants and all animals) or prokaryotes (of which bacteria are the best-known examples). Prokaryotes and eukaryotes have different cellular structures, the main difference being that eukaryotic cells have a distinct nucleus, containing DNA surrounded by a nuclear membrane, whereas in prokaryotes the DNA floats free in the cell and there is no separate nucleus.

recapitulation Controversial theory according to which an organism, during its development, 'climbs up its evolutionary tree'. The successive developmental stages of an organism would then resemble the successive adult ancestral stages in its evolutionary past. If true, the theory would provide a powerful method of reconstructing phylogenies; but many exceptions to it are known. The degree of generality of the theory has been much discussed but remains uncertain. The evolutionary form of *von Baer's law* should be distinguished: according to that law, successive developmental stages are successively more *derived character states*; whereas according to

Glossary

the theory of recapitulation they are successively more derived adult character states. An early embryo, according to the theory of recapitulation, represents an ancestral adult form; whereas according to von Baer's law it is not an ancestral adult but an undifferentiated form, the character states of which (or some of them) are relatively ancestral compared with later developmental stages.

sister species, sister groups A pair of species (or group of species) that share a more recent common ancestor with each other than with any third species.

stem species The common ancestral species of a group of descendant species.

subjective character choice, subjective classification Choice of characters, and therefore classification, without any reference to a theoretical principle. Contrast *objective classification*.

substance and accident In Aristotelian classification, species are thought to possess certain characters, called its substance, by virtue of which they are what they are. If the substance of a species is changed it becomes something essentially different. The substance of a species X is 'what-it-is-to-be' an X. Characters of a species other than its substance are called accidental, and the accidental characters of a species can change without altering the kind of species it is. For example, feathers might be the substance of the group 'birds', whereas it might be accidental whether a bird was perched in a tree or walking on the ground.

successive approximation A common scientific method whereby a theory is first tested and then, if it is confirmed, its validity is tentatively assumed in further tests. If the further tests are successful, even more confidence is placed in the original theory; but if the tests fail, the original theory will be suspected too. Often mistaken for argument in a vicious circle.

successive division In Aristotelian classification smaller subgroups are defined by division of the characters used to define the larger groups that contain them; the procedure is called successive division. Groups are not usually defined in this way in modern biology: the group Mammalia, for example, is part of the larger group Chordata; but characters, such as the possession of lactation, that define Mammalia, are not anatomical divisions of the characters, such as the possession of a notochord, that define the Chordata.

teleological classification Classification of groups by their shared purposes, or functions, in life – where purpose can be identified with adaptation. An imperfectly worked-out, occasionally suggested, theoretically possible principle of classification that differs from the two main such principles, *phenetic* and *phylogenetic classification*.

Evolution and Classification

transformed cladism Classification by means of shared *derived characters* (as in *cladism*) but without the assumption that the groups so defined are *phylogenetic*. Chapters 5-7 discuss possible reasons why one might wish to classify cladistically without assuming that the groups were evolutionary.

tree The *phylogeny* of a group of species. A rooted tree is a phylogeny in which the most ancestral species of the group has been recognized; an unrooted tree is one in which the *sister-species* relations are known but the direction of evolution is not, because it is not known which species is the most ancestral (see Figures 4.13 and 4.14).

uniformitarianism Philosophical doctrine that an observed process can also operate in unobserved times and places. It is the justification of all scientific extrapolation. Microevolution, for instance, has been observed, which suggests that all life has evolved from a common ancestor because the same process could have operated over larger time spans to produce larger changes. The argument uses the principle of uniformitarianism, but in a manner not to be confused with the much stricter theory of 'phyletic gradualism', according to which evolution not only has taken place when no one was looking, but also then took place always at a constant rate. Uniformitarianism, as here intended, suggests only that the same general process could have taken place, not that it always proceeded at the same rate.

von Baer's law Von Baer enunciated more than one embryological law; but only one of them is referred to in this book, the one which states that 'the general features of a large group of animals appear earlier in the embryo than the special features'. The characters of larger groups can therefore be discovered by looking at the earlier embryonic forms of its members. The characters of larger groups are likely to be relatively more *ancestral* than the characters of smaller subgroups. In an evolutionary interpretation, therefore, the law asserts that ancestral character states precede the derived ones in embryonic development; it thus gives a criterion by which to distinguish ancestral from derived character states (the *embryological criterion*). A corollary of the law is that the members of a group should resemble each other more closely at earlier developmental stages than as adults. The scope of the law is controversial.

weighting of characters Relative importance attached to different characters in classification. The characters given greater importance are said to be more heavily weighted. Characters can be numerically weighted in any manner by means of suitable adjustments in a *cluster statistic.*

Index

absent characters: criticized by Aristotle, 102-3
  liable to be convergent, 94-5
  not ancestral, 94
  a disastrous confusion, 95-7, 146-8
  do not justify transformed cladism, 91, 93-7
  synopsis, 15-16
adaptation, 5, 6, 125-37
  criterion of convergence, 21-5
  glossed, 183
Agassiz, L.:
  idealist morphology of, 107-11, 114
  not an ancestral cladist, 110
  mentioned, 16, 98
ambiguity, in classificatory principle, 6, 7
analogy, convergent character, 12, 20
  does not define monophyletic groups, 19-20
  defines artificial groups, 20
  recognition of, theoretically by functional criterion, 21-5, 128-30
  observationally by non-congruence, 25-8
  relative merits of techniques, 28-9
  loss convergences as, 94
  pre-Darwinian meaning, 111, 113
  glossed, 183
  mentioned, 78
Anatidae: Lorenz's behavioural classification of, 23
anatomie transcendente, 107
ancestors: search for, 10-11
  recognition of, 141-6
  even though said to be unrecognizable, 146-9
  cannot be found in modern samples, 138-9
  representation in cladism, 138-41
  unscientific, 146-9
  synopsis, 15
ancestral characters: meaning of, 54
  recognition of, 60-77
  can cause non-congruence, 54-6, 77
  define paraphyletic groups, 93
  absent, 93
  not absent, 94
  glossed, 183
  mentioned, 3, 14, 17
ancestral hypotheses: their confirmation, 141-2
  falsification, 142-3, 147
  and scientific imperfections, 147
Ancient Scottish Constitution, 114
apes: canine teeth of, 132-3
apomorphy, 184
Aquinas, St Thomas, 100, 108
Arachnida, 78, 130
archetype, 111; *see also* idealism, plan
Aristotle: on classification, 99-103
  influences Linnaeus and Cuvier, 103-6
  spurious use by transformed cladism, 95, 102
  not an ancestral cladist, 102-3
Arkansas: trial at, 117
Arthropoda: character disagreements in, 78
  polyphyly of? 128-30
artificial classification: explained, 7-8
  defined by convergent characters, 20
  imperfect distinction from natural classification, 44
  see also natural classification natural group

Index

glossed, 184
artificial selection, 119
average-neighbour statistic, 40
Avicenna, 100
axolotl: comfounds embryological criterion, 68

baboon, 54
bacteria gene transfer in, 49-51
Baer, K. von: his embryological law, 66-8, 90
  glossed, 192, *see also* embryological criterion; his idealist morphology, 107
  not an ancestral cladist, 110-12
  mentioned, 16, 95, 98
Balme, D. M.: on Aristotle, 100-2
barnacle, 4-5
birds, 5, 44, 113
  monophyletic, 29-30, 31, 59
  defined negatively by von Baer, 111
  teleologically by Mayr, 32-3
  divisions of, 94-5
Bock, W. J.: prefers evolutionary classification, 32
  and functional criterion, 127, 131, 134, 136
  dislikes outgroup comparison, 64, 124
butterfly, 14

Cain, A. J.: on adaptation, 24, 25, 126
  on circular arguments, in phylogenetic reconstruction, 35, 85
  and in testing evolution, 122-3
  opposes parsimony, 63
  on phenetic classification, 83
  defines MCD, 41
  on Cuvier, 103n, 105
  on Linnaeus, 103
  on Owen, 107
canine tooth: of humans, 132-3
Carson, H. L.: on Hawaiian *Drosophila*, 71-5
cave animals: vestigial organs in, 131-2
characters, classificatory: defined, 1, 77-8
  kinds of, 2-3
  numbers of, 2-3
  disagreement of, 2, 60-1, 77-83, 91, 185
  do not define evolutionary groups, 57-8
  weighting of, 35
character choice: necessary in classification, 2-3

Charig, A.: prefers evolutionary classification, 32
  on transformed cladism, 87-8, 93, 115-6
  on 'overall', 39n
chemical pigments: lepidopteran, 121-2
chromosomal inversions, used in phylogenetic reconstruction: of *Drosophila*, 71-5
  of great apes, 75-6
circular arguments: in tests of evolution, 122-4
  mentioned, 16
  in phylogenetic reconstruction, 27-8, 35, 84
  in outgroup comparison, 64-6
  in phenetic classification, 43
  in controversy, 124
cladism: a classificatory school, 6
  nature of, 46-85 (synopsis, 14; glossed, 184)
  its classifications, 55-7
  its naturalness, 58, 77
  opposes essentialism, 58
  prefers dichotomous classification, 150-5
  requires truth of evolution, 86, 90, 114
  tests evolution, 121
  representation of ancestors, 138-41
  best system of classification, 83-5, 86, 114
  wide support for, 85
  its dislike of evolutionary classification, 34
  effect on that school, 29, 32
  not a form of phenetic classification, 84-5
  the goal of numerical taxonomy, 45
  see also transformed cladism
cladogram: cladistic classification, 56, 89, 158-62, glossed, 184
  should be dichotomous? 150-7, 185
  synopsis, 15
classification: artificial and natural, 7-9
  objective and subjective, 3-7
  stability of, 3
  relation to evolutionary trees, 29-30
  tests evolution, 117-24
  evolutionary, 3-5, 29-30, *see also* evolutionary classification;
  phenetic, 3-5, 36-7, *see also* phenetic classification
  phylogenetic, 3-5, 55-7, *see also* cladism; teleological, 3, 5-6, *see also* teleological

Index

classification; of Arthropoda, 78, 128-30
  of barnacles, 4-5
  of birds, 94-5
  of Lepidoptera, 121
  of Mammalia, 20, 71, 105-6, 144-6
  of Polyzoa, 26
cluster statistic, 13; explained, 37; and glossed, 184
  multiplicity and conflicts of, 39-43
coelom, 26 glossed, 184
comparative method: investigates adaptation, 24-5, 128, 133-5
compatibility analysis, 82-3
congruence: glossed, 185
convergence, 4-5; defines artificial, 20
  not monophyletic groups, 19-20
  not represented in evolutionary classifications, 4-5, 30, 33-4
  represented in phenetic classifications, 4-5, 37, 91-2
  absent characters liable to, 94-5, 128
  recognition of, theoretically, by functional criterion, 21-5, 128-30
  observationally, by non-congruence, 25-8
  causes disagreement of characters, 77-8, 92, 126
  between marsupials and placentals, 20
  mentioned, 12, 17, 121
cow, 54
crab, 4
Cracraft, J.: on ancestors, 147
  on cladism, 88, 121
  on the functional criterion, 76, 127-8, 135-6
  on information, 84
  on pattern and process, 158, 161
  on ratites, 94
creationism: its forms, 116-7
  influence, 10, 122
  and errors, 116-24 *passim*
Crick, F. H. C.: on genetic code, 120
crocodile, 5, 29-30, 44, 54, 59
Crompton, A. W. (and F. W. Jenkins): on mammalian phylogeny, 33, 144-6
Crowson, R., 8, 11, 85
Crustacea, 78, 130
Cuvier, G.: classifications of, 103-6
  an ancestral cladist? 87
  no, 106
  mentioned, 16, 21, 98, 99, 100, 107, 113
cynodont, 71, 144-5
cytochrome *c*, sequences of, 79-81

Darwin, C., 21, 24, 29, 77, 106, 109-10, 122
Darwinian yoke, 18, 114
De Beer, Sir G. on embryological criterion, 67
  on ratites, 94-5
definition, of evolutionary groups, 57-8
derived characters: meaning of, 54
  glossed, 185
  recognition of, 60-77
  by outgroup comparison, 61-6
  embryologically, 66-8
  palaeontologically, 68-9
  disagreements of, not expected in theory, 58
  but expected in practice, 60-1
  how dealt with, 77-83
  needed in cladism, 54-5
  mentioned, 3, 14, 16, 17
developmental characters, *see* embryological criterion
dichotomous classification criticized by Aristotle, 102-3
  in cladism, 150-7
  mentioned, 185
differentia, Aristotelian term for defining character, 100-3
Dipnoi, 143
distance, phenetic, 36-7, 41
divergence, evolutionary: differential, 4-5, 12-13
  represented in evolutionary classification, 29-33
  but not in cladism, 58-60
Dobzhansky, T.: on biological species concept, 11
  on numerical taxonomy, 38
Dollo's law: in functional criterion, 135
Drosophila, Hawaiian: phylogeny of, 73-5
ducks: Lorenz's behavioural classification of, 23

ecological genetics, 119, 128
Ectoprocta, 26
Eldredge, N., 60, 69, 84, 124, 158-9
embryological criterion, of homology: golden age of, 21-2, 23
  mentioned, 17
  of character polarity, 66-8
  used by transformed cladists, 89-90
  mentioned, 76
  glossed, 185
Entoprocta, 26
Escherichia coli: gene transfer in, 49
  genetic code of, 119-20

195

Index

essentialism: of Aristotle, 99-103
  Aristotelian
  of Linnaeus, 103
  of Cuvier, 103-6
  teleological, 99-105
  theological, 99-100, 108
  does not justify transformed cladism, 103, 106, 113
  refuted by evolution, 57-8
  glossed, 186
  mentioned, 16, 99
evolution, test of: by classification, 117-8, 121-4
  mentioned, 15, 16
  by other means, 118-20
evolutionary classification: a school, 6-7
  its nature, 19-34 (synopsis, 12-13, glossed, 186)
  its justification, 19-20, 30-4, 83-4
  its techniques, 20-30
  would define monophyletic groups, 19
  on its own terms, 30-1
  if it can recognize them, 20-9
  phenetic element in, 29, 32
  subjectivity of, 43-4
  rejection of, 83-5
  relation to Cuvier, 104-5
  and to idealism, 111
  information content of, 34, 83-4
  humanity of, 34
  standard works, of, 19
evolutionary relations: discovery of, 20-9, 53-83
exclusion, principle of: cladistic use of, 76
  its absurdity, 77

Farris, J. S.: on aim of numerical taxonomy, 45
  on ancestors, 146-7
  on parsimony, 64, 82
Felsenstein, J.: on parsimony, 61
  on Hennig's dilemma, 78-83
fins, 3, 54, 93, 96
fish: defined by ancestral homologies, 93, 96, 97
  or teleologically, 33
  cladistic dismantling of, 34
  pre-Darwinian recognition of, 110-1
Fisher, D. C.: on *Limulus*, 128, 131
  on pattern and process, 161
Ford, E. B.: on lepidopteran pigments, 121-2
  mentioned, 128
Forey, P. L.: prefers transformed cladism, 88
  on ancestors, 146
fossil record *see* palaeontology
Foulds, L. R.: tests evolution, 122
frozen accident, 120
function: explains character correlation, 20, 104
functional analysis, 125-37
  glossed, 186
  mentioned, 12, 17
functional criterion of homology, 94-5, 125-30
  mentioned, 17
  of character polarity, 131-7
Fürbringer, M., 94
furthest-neighbour cluster statistic, 40

Gardiner, B. G.: prefers transformed cladism, 88
  his controversial classification, 143-4
Garstang, W., 24
Genesis: evidence against, 117-23 *passim*
genetic code: universality of, 119-20
  glossed, 186-7
geology:
  evidence for evolution, 117, 118-19
  roots trees, 73
Ghiselin, M. T.: prefers evolutionary cladism, 85
  opposes essentialism, 57
  and transformed cladism, 57, 93, 95
  distinguishes history from criticism, 99
  mentioned, 8, 18
God: a transformed cladist? 108-9
  unfathomable thoughts of, 109, 110, 113-14
Goodman, M., 24
Gould, S. J.: prefers evolutionary classification, 32
  on embryological criterion, 21, 66-7
  on macroevolution, 33
  mentioned, 90
grade, 32-3
  glossed, 187
  see also teleological classification
Gutmann, W. F.: on functional criterion, 131, 134

Haeckel, E., 23, 90, 98
Halstead, L. B., 10, 60, 144
  prefers evolutionary classification, 32, 34
Harper, C. W.: on cladistic sister groups, 139, 141

196

Index

Hawaiian *Drosophila*: phylogeny of, 73–5
Hendy, M. D.: tests evolution, 122
  on trees, 161
Hennig, W.: justifies cladism, 46–53, 93
  mentioned, 60
  on cladism, 9, 14–15, 27, 54, 64, 84, 86
  supporters of, 85
  not a transformed cladist, 9–10, 88–90
  defines monophyletic, 30, 32, 52
  on dichotomous classification, 150–5
Hennig's dilemma, 77–83
holophyletic group, synonym of cladistic monophyly, 53
  glossed, 187
homology, ancestral character similarity, 12, 20–9
  meanings of, 30
  pre-Darwinian meaning, 107, 111, 113, 117
  glossed, 187
  recognition of, theoretically, by functional criterion, 21–5, 94, 128–30
  observationally, by non-congruence, 25–8
  relative merits of the techniques, 28–9
  used in evolutionary classification, 29–30
  ancestral and derived forms distinguished, 30, 53–7
  test evolution, 117–18; universal, 119–20
  mentioned, 12, 17, 78, 103–4
Hull, D. L.: on philosophy, 10, 147–8
  on circular argument, 123
  in phylogenetic reconstruction, 27–8
  in outgroup comparison, 64
  in tests of evolution, 116
  on essentialism, 11, 58
  on phenetic classification, 18, 162
  on cladism, 76, 159
  on dichotomous classification, 150, 152, 157
  mentioned, 29, 106
Humphries, C. J.: on the functional criterion, 127
  on pattern and process, 158
Huxley, J.: 'new systematics' of, 11
  defines grades, 32
  on impossibility of evolutionary classification, 85
hypotheses: grounds for preferring, 156

idealism: divine, 107
  distinguished from analogical essentialism, 108
  secular, 107–8
  glossed, 187
information content: of evolutionary classification, 34, 83–4
  of phenetic classification, 83
  a dangerous criterion, 84, 156
  greater in dichotomous classification? 155–6
  mentioned, 14, 15
inversions, chromosomal, 71–5
  glossed, 187

Janvier, P.: on transformed cladism, 81, 91, 115
Johnson, L. A. S.: criticizes numerical phenetics, 39–43
  mentioned, 13, 18, 124
  prefers evolutionary classification, 32, 83
justification: meaning of, 2
  requirements of, 7
  of evolutionary classification, 12, 19–20
  of phenetic classification, 13, 35–6
  of cladism, 14, 46–51
  of transformed cladism, 15–17, 91–124

Kemp, T.: on mammalian phylogeny, 33, 70, 143–6, 148
Kimura, M., 22

Lande, R., 94
Lankester, E. R.: on embryological criterion, 21–2, 23
Le Gros Clark, W. E.: on direction of evolution, 132–3
Lepidoptera: classification of, 66–7, 121
limbs, 3, 93, 96
limpet, 4–5
Limulus: Fisher on, 128, 131
Linnaeus, C.: classification of, 103
  not an ancestral cladist, 106
  mentioned, 16, 98, 99, 100, 107, 113
Linnaen hierarchy, 1
  glossed, 187
lizard, 5, 29–30, 44
Lorenz, K.: on homology, 23
loss convergence, 94–5, 128
lungfish, 143
Lyell, C., 117, 118

MacLeay, W. S.: theological idealism of, 107
  not an ancestral cladist, 110
  mentioned, 98

197

Index

Mammalia: phylogeny of, 71, 144-6
  defined monophyletically, 31, 33
  teleologically, 33
  Cuvier's classification of, 105-6
  mammal-like reptiles unrooted tree of, 70-71
Manton, S., 78, 128-30, 136-7
Marsupialia, 20, 106
Marxism: sniffed by Dr Halstead, 10
Maynard Smith, J., 24n, 119, 128
Mayr, E.: prefers evolutionary classification, 19, 32-3, 83-4
  prefers teleological classification, 32-3
  mentioned, 8, 11, 20, 29, 85, 154
  criticizes phenetic classification, 38
  and cladism, 59-60, 157
  and parsimony, 83
  his hopeless definitions of monophyly, 30, 31, 52-3
  on loss convergences, 94
  defines 'character', 1
Menenius Agrippa, 18
minimum evolution tree, 79, 82-3
  glossed, 187-8
molecular characters and molecular taxonomy inspiration of, 22
  of cytochrome *c*, 78-82
molecular clock, 22
  glossed, 188
Mollusca, 5
monophyletic groups: four meanings of, 30-2, 51-3
  defined by homologies, 29-30
  defined by shared derived characters, 55
  exist independently of the characters used to recognize them, 57-8
  the aim of evolutionary classification, 19-20, 32
  and cladism, 51
  glossed, 188
morphological idealism, 106-13
  does not justify transformed cladism, 110-11, 113-14
morphological similarity, *see* phenetic similarity
morphology: glossed, 188
moth, 4
Müller, F., 23

natural classification: meanings of, 7-9, 98, 107
  defined, 8
  an imprecise concept, 44, 59, 104, 108
  due to evolution, 8-9, 12, 19, 58
  tests evolution, 117-18, 121-2
  glossed, 188
natural groups: due to evolution, 8-9, 12, 19, 58
natural selection: gives purpose to living things, 6, 113
  helps to recognize homologies, 21-5, 125-30
  and derived characters, 131-7
  mentioned, 12, 17
natural theology, 106-7, 110
Nature: editorial *terribilità,* 10
Naturphilosophie, 107
Naudin, C., 29
nauplius, 5
nearest-neighbour cluster statistic, 37
negative characters, *see* absent characters
Nelson, G.: prefers transformed cladism, 87-93
  on absences, 93-9 *passim;* on ancestors, 147
  on Aristotle, 95, 102-3
  on dichotomy, 157
  on embryological criterion, 90
  on information, 156
  on outgroup comparison, 64
neutrality, selective: a precarious evolutionary philosophy, 24
New Systematics: interests of, 11
niche, Eltonian, 6
nomenclature, 11
non-congruency: implicitly glossed, 185
  see also characters, classificatory: disagreement of
Norman yoke, 114
numerical phenetics, *see* phenetic classification
numerical taxonomy, 13, 27, 28
  a positivist enthusiasm, 17-18, 124
  glossed, 188; *see also* phenetic classification

objective classification: defined, 3
  principles of, 6-7
  cladism as, 47-9, 53, 59-60
  glossed, 188
  mentioned, 16
  see also subjective classification
Ockham, William of, 82
Onychophora, 78, 130
operational taxonomic unit, 36, 86
operationism: glossed, 188-9
ophiacodontid, 70
outgroup comparison, 61-6
  parsimonious, 61-4
  not circular, 64-7

Index

roots trees, 70, 71, 73
  glossed, 189
  mentioned, 76
'overall', 39n
Owen, R.: his idealist morphology, 107, 110-13
  theological, 107
  not an ancestral cladist, 110-11
  mentioned, 16, 98

palaeontology: used to distinguish derived characters, 60, 68-9, 76
  to root trees, 71
  to test evolution, 118-19
  disciplinary status of, 10-11
Paley, W., 106
Pantin's student: lament of, 76
paraphyletic groups: explained, 52-3
  fish as, 96
  reptiles as, 96
  glossed, 189
parsimony: meaning of, 64
  assumed in outgroup comparison, 61-4
  resolves Hennig's dilemma, 79, 82-3, 126
  valid justification, 61, 82
  and invalid, 82
  glossed, 189
pattern cladism, *see* transformed cladism
pattern-process distinction, 14, 16-17, 18
  affronted by functional criterion, 126, 161
  and much else, 158-62
  erroneous, 159, 161-2
  glossed, 189
Patterson, C.: prefers transformed cladism, 87-90, 92, 95-7
  on absences, 93, 95
  on ancestors, 148
  on Aristotle, 95, 102-3
  on embryological criterion, 66
  on history, 98
  on homology, 30
  on outgroup comparison, 64
  on palaeontology, 76
  on tests of evolution, 115
Penny, D.: tests evolution, 122
  on trees, 161
Peripatus, 78, 130
phenetic classification: a school, 6
  nature of, 3-5, 35-45 (synopsis, 13; glossed, 189-90)
  justification of, 35-6

techniques of, 36-8
  repeatability of, 36, 38-9
  subjectivity of, 43-5, 91-3, 110
  rejection of, 83-5
  not evolutionary, 38, 86
  but would test evolution, 122-3
  dislikes evolutionary classification, 34, 35
  and cladism, 84-5
  resembles idealism, 108
  and transformed cladism, 91-2, 162
  mentioned, 7, 29
phenetic similarity: represented in classification, 30-3
  not represented in cladism, 58-60
  measurement of, 36-7, 39-43
  no natural hierarchy of, 43, 46-9, 59, 85, 104
Philo, 109, 110
philosophy: meaning of, 2
phylogenetic classification: nature of, 3-5
  justification of, 46-53
  techniques of, 53-83
  glossed, 190
  mentioned, 7
  see also cladism
phylogenetic reconstruction: methods of, 20-9, 83; *see also* cladism
phylogeny meaning of, 46n
  true natural hierarchy of, 47-9
  even in prokaryotes, 50-1
  biological interest of, 85
  glossed, 190
placentals, 20
plan, ideal: of groups, 106-13
Platnick, N.: prefers transformed cladism, 18, 87-92, 162
  on absences, 93-7
  on ancestors, 147
  on Aristotle, 95, 102-3
  on dichotomy, 154-7
  on history, 98
  on tests of evolution, 115
plesiomorphy, 184
polemical literature: irrelevance of, 10-11
polyphyletic groups, 52-3
  artificial, 20
  defined by convergent characters, 19-20 (e.g. coelom? 26)
  not allowed in evolutionary classification, 19-20
  nor cladism, 52-4
  monophyletic groups as, 31
  Arthropoda as? 128-30

199

Index

Mammalia as? 31
 glossed, 190
Polyzoa, 26
Popper, K.: his philosophy, misapplied:
 to ancestral hypotheses, 148
 to choice of cluster statistic, 92-3
 to 'information', 156
 to neutralism, 24n
 his self-confessed irrelevance, 10
 mentioned, 14, 122
practical possibility: not a criterion of objectivity, nor a justification, 16, 90, 98
precopulatory mate guarding: evolutionary polarity of, 133-5
pre-Darwinian classification: no longer acceptable, 18
 specious justification of transformed cladism, 91, 98-114
 mentioned, 15, 16
primates, 110
principle, of character choice:
 explained, 3
 requirements of, 6-7
 practicality, 7
 nonambiguity, 7
 naturalness, 3, 7
privations, Aristotelian term for absent characters (q.v.)
Probainognathus: ancestral mammal? 144-6
progress, evolutionary, 131
prokaryotes: gene transfer among, 49-50
 a worry for cladism? 49
 no, 50-1
 glossed, 190

ratites: monophyletic? 94-5
recapitulation theory of, 22
 glossed, 190
reciprocal illumination, 27, 64
repeatability: of numerical phenetic classification, 36, 38-9
Reptilia: definition of,
 monophyletically, 31, 32, 52
 teleologically, 33
 phenetically, 44
 by ancestral homologies, 58-9, 96
 pre-Darwinian recognition of, 110-11, 114
revolutions, in philosophical taxonomy:
 their course, 17-18
 and myths, 18, 114
R-factors, in bacteria, 49, 51
Rhipidistia, 143

Rieppel, O.: prefers transformed cladism, 87-8, 115
Rodentia: their exemplary gnawing apparatus, 25
Rosen, D. E.: prefers transformed cladism, 88
 (*et al.*) controversial classification, 143-4
'scenario', 158-62
schools, classificatory, 6-7, 9-10
 philosophical not religious, 12
Severtzoff, A. N.: on functional criterion, 131
Shigella: gene transfer in, 49
signals,
 animal: use in classification, 23, 24
Simpson, G. G.: prefers evolutionary classification, 19, 24, 29, 32
 prefers teleological classification, 32-3
 on biological species concept, 11
 on dichotomy, 154, 157
 on monophyly, 31, 32
 on naturalness, 8-9
 mentioned, 20
single-linkage cluster statistic, 37, 39
sister groups and sister species: defined, 138
 strange cladistic meaning of, 140-1, 146
 glossed, 191
Sneath, P. H. A.: prefers numerical phenetics, 11, 35-45
 criticizes evolutionary classification, 27-8
 and cladism, 47, 51, 64, 84-5
 on monophyly, 31
Sokal, R. R.: prefers numerical phenetics, 11, 35-45
 criticizes evolutionary classification, 27-8
 and cladism, 47, 51, 84-5
 on monophyly, 31
speciation: dichotomous? 150-7
synopsis, 15
species concept, 11-12
stability, of classification, 3, 8
stem species, 147
 glossed, 191
Stevens, P. F.: on cladism, 61, 68, 87, 88
subjective classification: defined, 3
 a property of phenetic classification, 13, 39-45
 transformed cladism too, 15, 91-3, 124
 glossed, 191
substance, 99-103; glossed, 191

200

see also essentialism, of Aristotle
successive approximation: in outgroup comparison, 64-6
  recognition of homologies, 27-8
  tests of evolution, 123-4
  glossed, 191
successive division: in Aristotelian classification, 101-2
  not necessary elsewhere, 95
  glossed, 95
synapomorphy, 185
Szalay, F. S.: prefers evolutionary classification, 32
  on functional criterion, 133

taxonomic controversy: fundamental source of, 2
taxonomists: apotheosis of, 109
  habits of, 1-2
taxonomy, *see* classification
teleological classification: defined, 3, 5-6
  its essentialist form, 99-105
  preferred by Aristotle, 99-103
  and Linnaeus, 103
  and Cuvier, 103-6
  in evolutionary classification, 32-4, 43
  its informality, 33-4; 100, 113
  glossed, 191
  mentioned, 125
Tetrapoda, 96-7
theological principle: in essentialism, 99-100
  in idealism, 107-10
  its impracticality, 109-10
  mentioned, 16
therapsid, 71
Thompson, J. V.: on barnacles, 5
Thompson, K. S.: flirts with transformed cladism, 115
transformed cladism: a school, 9-10
  nature of, 86-91 (synopsis, 14; glossed, 192)
  dislikes evolution, 9, 10, 87-8
  controversial, 9-11
  even terrorist, 143
  a version of phenetics, 91-3
  a positivist enthusiasm, 17-18, 124
  techniques of, 9, 89-90
  affronted by functional criterion, 125-7

  practical possibility of, 16, 89, 90
  justification of, 9, 91-124 (synopsis, 15-16), non-existent? 91-3
  by absent characters, 93-7
  by history, 98-114
  seeks comfort in Aristotle, 95, 102-3
  and pre-Darwinian idealism, 98, 107, 108
  all in vain, 110-11, 113-14
  justification as test of evolution, 115-24
  subjectivity of, 91-3, 97, 114, 124
  rejection of, 124
  prefers dichotomous classification, 151
  for no good reason, 155-7
tree, 158-62
  glossed, 192
  see also phylogeny
Tritylodontidae, 144-6

uniformitarianism, 119
  glossed, 192
universal homologies, 119-20
unrooted tree: meaning of, 70
  discovery of, 70-2
  of *Drosophila*, 73-5
  of mammal-like reptiles, 71

Vertebrata: von Baer's classification, 112
  Cuvier's, 105
  mentioned, 94
vestigial organs: evolutionarily derived, 131-3
  planned, 107
viruses: no difficulty for cladism, 51

wasp, 4
Wasserman, M.: on chromosomal inversions, 73
weighting, 3, 35
  disliked by numerical taxonomy, 35-6, 45
  glossed, 192
Weismann, A.: on embryological criterion, 66-7
Williamson, M.: on Hawaiian *Drosophila*, 73-5
Wilson, E. O.: his consistency test, 8
Witan, 114